An OPUS book

A Historical Introduction to the Philosophy of Science

A Historical Introduction
to the
Philosophy of Science

Third edition, revised and enlarged

JOHN LOSEE

Oxford · New York
OXFORD UNIVERSITY PRESS

Oxford University Press, Walton Street, Oxford OX2 6DP

Oxford New York
Athens Auckland Bangkok Bombay
Calcutta Cape Town Dar es Salaam Delhi
Florence Hong Kong Istanbul Karachi
Kuala Lumpur Madras Madrid Melbourne
Mexico City Nairobi Paris Singapore
Taipei Tokyo Toronto

and associated companies in
Berlin Ibadan

Oxford is a trade mark of Oxford University Press

First published as an Oxford University Press
paperback 1972 and simultaneously in a hardback edition
Paperback reprinted 1977
Second edition in paperback and hardback 1980
Paperback reprinted 1984, 1985
Third edition in paperback 1993

British Library Cataloguing in Publication Data
Data available

Library of Congress Cataloging in Publication Data
Losee, John.
    A historical introduction to the philosophy of science / John
Losee.—3rd ed., rev. and enlarged.
        p. · cm. "An OPUS book"—Cover.
    1. Science—Philosophy—History. I. Title.
    501—dc20 Q174.8.L67 1993 93-22328
    ISBN 0-19-289247-9

10 9 8 7 6 5 4

Printed in Great Britain by
Biddles Ltd
Guildford and King's Lynn

Preface

This book is a historical sketch of the development of views about scientific method. Its emphasis is on developments prior to 1940. No attempt has been made to reproduce the contemporary spectrum of positions on the philosophy of science. My purpose has been exposition rather than criticism, and I have endeavoured to abstain from passing judgement on the achievements of the great philosophers of science.

It is my hope that this book may be of interest both to students of the philosophy of science and to students of the history of science. If, on reading this book, a few such students are encouraged to consult some of the works listed in the Bibliography at the end of the book, I shall consider my effort to have been well spent.

I have received numerous helpful suggestions from Gerd Buchdahl, George Clark, and Rom Harré in the preparation of this volume. I am most grateful, both for their encouragement, and for their criticism. Of course, responsibility for what has emerged is mine alone.

Lafayette College
July 1971

Preface to the Second Edition

The discussion of post-Second-World-War developments has been reorganized and expanded in the second edition. There are new chapters on the Logical Reconstructionism of Carnap, Hempel, and Nagel; the critical reaction to this orientation; and the alternative approaches of Kuhn, Lakatos, and Laudan.

August 1979

Preface to the Third Edition

The third edition includes new material on theories of scientific progress, causal explanation, Bayesian confirmation theory, scientific realism, and alternatives to prescriptive philosophy of science.

September 1992

Contents

Introduction

A decision on the scope of the philosophy of science is a precondition for writing about its history. Unfortunately, philosophers and scientists are not in agreement on the nature of the philosophy of science. Even practising philosophers of science often disagree about the proper subject-matter of their discipline. An example of this lack of agreement is the exchange between Stephen Toulmin and Ernest Nagel on whether philosophy of science should be a study of scientific achievement *in vivo*, or a study of problems of explanation and confirmation as reformulated in the terms of deductive logic.[1] To establish a basis for the subsequent historical survey, it will be helpful to sketch four viewpoints on the philosophy of science.

One view is that the philosophy of science is the formulation of world-views that are consistent with, and in some sense based on, important scientific theories. On this view, it is the task of the philosopher of science to elaborate the broader implications of science. This may take the form of speculation about ontological categories to be used in speaking about "being-as-such". Thus Alfred North Whitehead urged that recent developments in physics require that the categories 'substance' and 'attribute' be replaced by the categories 'process' and 'influence'.[2] Or it may take the form of pronouncements about the implications of scientific theories for the evaluation of human behaviour, as in Social Darwinism and the theory of ethical relativity. The present study is not concerned with "philosophy of science" in this sense.

A second view is that the philosophy of science is an exposition of the presuppositions and predispositions of scientists. The philosopher of science may point out that scientists presuppose that nature is not capricious, and that there exist in nature regularities of sufficiently low complexity to be accessible to the investigator. In addition, he may uncover the preferences of scientists for deterministic rather than statistical laws, or for mechanistic rather than teleological explanations. This view tends to assimilate philosophy of science to sociology.

A third view is that the philosophy of science is a discipline in which the concepts and theories of the sciences are analysed and clarified. This is not a matter of giving a semi-popular exposition of the latest theories. It is, rather, a matter of becoming clear about the meaning of such terms as 'particle', 'wave', 'potential', and 'complex' in their scientific usage.

But as Gilbert Ryle has pointed out, there is something pretentious about this view of the philosophy of science—as if the scientist needed the philosopher of science to explain to him the meanings of scientific concepts.[3] There would seem to be two possibilities. Either the scientist does understand a concept that he uses, in which case no clarification is required. Or he does not, in which case he must inquire into the relations of that concept to other concepts and to operations of measurement. Such an inquiry is a typical scientific activity. No one would claim that each time a scientist conducts such an inquiry he is practising philosophy of science. At the very least, we must conclude that not every analysis of scientific concepts qualifies as philosophy of science. And yet it may be that certain types of conceptual analysis should be classified as part of the philosophy of science. This question will be left open, pending consideration of a fourth view of the philosophy of science.

A fourth view, which is the view adopted in this work, is that philosophy of science is a second-order criteriology. The philosopher of science seeks answers to such questions as:

1. What characteristics distinguish scientific inquiry from other types of investigation?
2. What procedures should scientists follow in investigating nature?
3. What conditions must be satisfied for a scientific explanation to be correct?
4. What is the cognitive status of scientific laws and principles?

To ask these questions is to assume a vantage-point one step removed from the practice of science itself. There is a distinction to be made between doing science and thinking about how science ought to be done. The analysis of scientific method is a second-order discipline, the subject-matter of which is the procedures and structures of the various sciences, viz.:

LEVEL	DISCIPLINE	SUBJECT-MATTER
2	Philosophy of Science	Analysis of the Procedures and Logic of Scientific Explanation
1	Science	Explanation of Facts
0		Facts

The fourth view of the philosophy of science incorporates certain aspects of the second and third views. For instance, inquiry into the predispositions of scientists may be relevant to the problem of evaluating scientific theories. This is particularly true for judgements about the completeness of explanations. Einstein, for example, insisted that statistical accounts of radioactive decay were incomplete. He maintained that a complete interpretation would enable predictions to be made of the behaviour of individual atoms.

In addition, analyses of the meanings of concepts may be relevant to the demarcation of scientific inquiry from other types of investigation. For instance, if it can be shown that a term is used in such a way that no means are provided to distinguish its correct application from incorrect application, then interpretations in which the concept is embedded may be excluded from the domain of science. Something like this took place in the case of the concept 'absolute simultaneity'.

The distinction which has been indicated between science and philosophy of science is not a sharp one. It is based on a difference of intent rather than a difference in subject-matter. Consider the question of the relative adequacy of Young's wave theory of light and Maxwell's electromagnetic theory. It is the scientist *qua* scientist who judges Maxwell's theory to be superior. And it is the philosopher of science (or the scientist *qua* philosopher of science) who investigates the general criteria of acceptability that are implied in judgements of this type. Clearly these activities interpenetrate. The scientist who is ignorant of precedents in the evaluation of theories is not likely to do an adequate job of evaluation himself. And the philosopher of science who is ignorant of scientific practice is not likely to make perceptive pronouncements on scientific method.

Recognition that the boundary-line between science and philosophy of science is not sharp is reflected in the choice of subject-matter for this historical survey. The primary source is what scientists and philosophers have said about scientific method. In some cases this is sufficient. It is possible to discuss the philosophies of science of Whewell and Mill, for example, exclusively in terms of what they have written about scientific method. In other cases, however, this is not sufficient. To present the philosophies of science of Galileo and Newton, it is necessary to strike a balance between what they have written about scientific method and their actual scientific practice.

Moreover, developments in science proper, especially the introduction of new types of interpretation, subsequently may provide grist for the mill of philosophers of science. It is for this reason that brief accounts have been included of the work of Euclid, Archimedes, and the classical atomists, among others.

Notes

[1] Stephen Toulmin, *Sci. Am.* 214, no. 2 (Feb. 1966), 129–33; 214, no. 4 (Apr. 1966), 9–11;

Ernest Nagel, *Sci. Am.* 214, no. 4 (Apr. 1966), 8–9.

[2] Whitehead himself did not use the term 'influence'. For his position on the relation of science and philosophy see, for example, his *Modes of Thought* (Cambridge: Cambridge University Press, 1938), 173–232.

[3] Gilbert Ryle, 'Systematically Misleading Expressions', in A. Flew, ed., *Essays on Logic and Language—First Series* (Oxford: Blackwell, 1951), 11–13.

1

Aristotle's Philosophy of Science

ARISTOTLE (384–322 BC) was born in Stagira in northern Greece. His father was physician to the Macedonian court. At the age of 17 Aristotle was sent to Athens to study at Plato's Academy. He was associated with the Academy for a period of twenty years. Upon Plato's death in 347 BC, and the subsequent election of the mathematically-oriented Speucippus to head the Academy, Aristotle chose to pursue his biological and philosophical studies in Asia Minor. In 342 BC he returned to Macedonia as tutor to Alexander the Great, a relationship which lasted two or three years.

By 335 BC Aristotle had returned to Athens and had established the Peripatetic School in the Lyceum. In the course of his teaching at the Lyceum, he discussed logic, epistemology, physics, biology, ethics, politics, and aesthetics. The works that have come to us from this period appear to be compilations of lecture notes rather than polished pieces intended for publication. They range from speculation about the attributes predicable of 'being-as-such' to encyclopedic presentations of data on natural history and the constitutions of Greek city-states. The *Posterior Analytics* is Aristotle's principal work on the philosophy of science. In addition, the *Physics* and the *Metaphysics* contain discussions of certain aspects of scientific method.

Aristotle left Athens after the death of Alexander in 323 BC, lest Athens "sin twice against philosophy". He died the following year.

Aristotle was the first philosopher of science. He created the discipline by analysing certain problems that arise in connection with scientific explanation.

Aristotle's Inductive–Deductive Method

Aristotle viewed scientific inquiry as a progression from observations to general principles and back to observations. He maintained that the scientist should induce explanatory principles from the phenomena to be explained, and then deduce statements about the phenomena from premises which include these principles. Aristotle's inductive–deductive procedure may be represented as follows:

Aristotle believed that scientific inquiry begins with knowledge that certain events occur, or that certain properties coexist. Scientific explanation is achieved only when statements about these events or properties are deduced from explanatory principles. Scientific explanation thus is a transition from knowledge of a fact (point (1) in the diagram above) to knowledge of the reasons for the fact (point (3)).

For instance, a scientist might apply the inductive–deductive procedure to a lunar eclipse in the following way. He begins with observation of the progressive darkening of the lunar surface. He then induces from this observation, and other observations, several general principles: that light travels in straight lines, that opaque bodies cast shadows, and that a particular configuration of two opaque bodies near a luminous body places one opaque body in the shadow of the other. From these general principles, and the condition that the earth and moon are opaque bodies, which, in this instance, have the required geometrical relationship to the luminous sun, he then deduces a statement about the lunar eclipse. He has progressed from factual knowledge that the moon's surface has darkened to an understanding of why this took place.

The Inductive Stage

According to Aristotle, every particular thing is a union of matter and form. Matter is what makes the particular a unique individual, and form is what makes the particular a member of a class of similar things. To specify the form of a particular is to specify the properties it shares with other particulars. For example, the form of a particular giraffe includes the property of having a four-chambered stomach.

Aristotle maintained that it is by induction that generalizations about forms are drawn from sense experience. He discussed two types of induction. The two types share the characteristic of proceeding from particular statements to general statements.

The first type of induction is simple enumeration, in which statements about individual objects or events are taken as the basis for a generalization about a species of which they are members. Or, at a higher level, statements about individual species are taken as a basis for a generalization about a genus.

Aristotle's First Type of Induction: Simple Enumeration

Premisses		Conclusion
what is observed to be true of several individuals	$\xrightarrow{\text{generalization}}$	what is presumed to be true of the species to which the individuals belong
what is observed to be true of several species	$\xrightarrow{\text{generalization}}$	what is presumed to be true of the genus to which the species belong

In an inductive argument by simple enumeration, the premisses and conclusion contain the same descriptive terms. A typical argument by simple enumeration has the form:

$$a_1 \text{ has property } P$$
$$a_2 \quad '' \quad\quad '' \quad P$$
$$a_3 \quad '' \quad\quad '' \quad P$$
$$\therefore \text{ All } a\text{'s have property } P.^*$$

---

* A double line between premisses and conclusion is used to indicate that the argument is an inductive one.

The second type of induction is a direct intuition of those general principles which are exemplified in phenomena. Intuitive induction is a matter of insight. It is an ability to see that which is "essential" in the data of sense experience. An example given by Aristotle is the case of a scientist who notices on several occasions that the bright side of the moon is turned toward the sun, and who concludes that the moon shines by reflected sunlight.[1]

The operation of intuitive induction is analogous to the operation of the "vision" of the taxonomist. The taxonomist is a scientist who has learned to "see" the generic attributes and *differentiae* of a specimen. There is a sense in which the taxonomist "sees more than" the untrained observer of the same specimen. The taxonomist knows what to look for. This is an ability which is achieved, if at all, only after extensive experience. It is probable that when Aristotle wrote about intuitive induction, this is the sort of "vision" he had in mind. Aristotle himself was a highly successful taxonomist who undertook to classify some 540 biological species.

The Deductive Stage

In the second stage of scientific inquiry, the generalizations reached by induction are used as premises for the deduction of statements about the initial observations. Aristotle placed an important restriction on the kinds of statements that can occur as premises and conclusions of deductive arguments in science. He allowed only those statements which assert that one class is included within, or is excluded from, a second class. If 'S' and 'P' are selected to stand for the two classes, the statements that Aristotle allowed are:

Type	Statement	Relation
A	All *S* are *P*	*S* wholly included in *P*
E	No *S* are *P*	*S* wholly excluded from *P*
I	Some *S* are *P*	*S* partially included in *P*
O	Some *S* are not *P*	*S* partially excluded from *P*

Aristotle held that type *A* is the most important of these four types. He believed that certain properties inhere essentially in the individuals of certain classes, and that statements of the form 'All *S* are *P*' reproduce the structure of these relations. Perhaps for this reason, Aristotle maintained that a proper scientific explanation should be given in terms of statements of this type. More specifically, he cited the syllogism in Barbara as the paradigm of scientific demonstration. This syllogism consists of *A*-type statements arranged in the following way:

All *M* are *P*.
All *S* are *M*.
∴ All *S* are *P*.

where *P*, *S*, and *M* are the major, minor, and middle terms of the syllogism.

Aristotle showed that this type of syllogism is valid. If it is true that every *S* is included in *M* and every *M* is included in *P*, it also must be true that every *S* is included in *P*. This is the case regardless of what classes are designated by '*S*', '*P*', and '*M*'. One of Aristotle's great achievements was to insist that the validity of an argument is determined solely by the *relationship* between premisses and conclusion.

Aristotle construed the deductive stage of scientific inquiry as the interposition of middle terms between the subject and predicate terms of the statement to be proved. For example, the statement 'All planets are bodies that shine steadily' may be deduced by selecting 'bodies near the earth' as middle term. In syllogistic form the proof is:

All bodies near the earth are bodies that shine steadily.
All planets are bodies near the earth.
∴ All planets are bodies that shine steadily.

Upon application of the deductive stage of scientific procedure, the scientist has advanced from knowledge of a fact about the planets to an understanding of why this fact is as it is.[2]

Empirical Requirements for Scientific Explanation

Aristotle recognized that a statement which predicates an attribute of a class term always can be deduced from more than one set of premisses. Different arguments result when different middle terms are selected, and some arguments are more satisfactory than others.

The previously given syllogism, for instance, is more satisfactory than the following:

> All stars are bodies that shine steadily.
> All planets are stars.
> ∴ All planets are bodies that shine steadily.

Both syllogisms have the same conclusion and the same logical form, but the syllogism immediately above has false premises. Aristotle insisted that the premises of a satisfactory explanation must be true. He thereby excluded from the class of satisfactory explanations those valid syllogisms that have true conclusions but false premises.

The requirement that the premises be true is one of four extra-logical requirements which Aristotle placed on the premises of scientific explanations. The other three requirements are that the premises must be indemonstrable, better known than the conclusion, and causes of the attribution made in the conclusion.[3]

Although Aristotle did state that the premises of every adequate scientific explanation ought to be indemonstrable, it is clear from the context of his presentation that he was concerned to insist only that there must be *some* principles within each science that cannot be deduced from more basic principles. The existence of some indemonstrable principles within a science is necessary in order to avoid an infinite regress in explanations. Consequently, not all knowledge within a science is susceptible to proof. Aristotle held that the most general laws of a science, and the definitions which stipulate the meanings of the attributes proper to that science, are indemonstrable.

The requirement that the premises be "better known than" the conclusion reflects Aristotle's belief that the general laws of a science ought to be self-evident. Aristotle knew that a deductive argument can convey no more information than is implied by its premises, and he insisted that the first principles of demonstration be at least as evident as the conclusions drawn from them.

The most important of the four requirements is that of causal relatedness. It is possible to construct valid syllogisms with true premises in such a way that the premises fail to state the cause of the attribution which is made in the conclusion. It is instructive to compare the following two syllogisms about ruminants, or cud-chewing animals:

Syllogism of the Reasoned Fact

All ruminants with four-chambered stomachs are
   animals with missing upper incisor teeth.
All oxen are ruminants with four-chambered stomachs.

∴ All oxen are animals with missing upper incisor teeth.

Syllogism of the Fact

All ruminants with cloven hoofs are animals with
   missing upper incisor teeth.
All oxen are ruminants with cloven hoofs.

∴ All oxen are animals with missing upper incisor teeth.

Aristotle would say that the premisses of the above syllogism of
the reasoned fact state the cause of the fact that oxen have missing
incisors in the upper jaw. The ability of ruminants to store partially
chewed food in one stomach chamber and to return it to the
mouth for further mastication explains why they do not need, and
do not have, incisors in the upper jaw. By contrast, the premisses
of the corresponding syllogism of the fact do not state the cause of
the missing upper incisors. Aristotle would say that the correlation
of hoof structure and jaw structure is an accidental one.

What is needed at this point is a criterion to distinguish causal
from accidental correlations. Aristotle recognized this need. He
suggested that in a causal relation the attribute (1) is true of every
instance of the subject, (2) is true of the subject precisely and not
as part of a larger whole, and (3) is "essential to" the subject.

Aristotle's criteria of causal relatedness leave much to be desired.
The first criterion may be applied to eliminate from the class of
causal relations any relation to which there are exceptions. But one
could establish a causal relation by applying this criterion only for
those cases in which the subject class can be enumerated completely.
However, the great majority of causal relations of interest to the
scientist have an open scope of predication. For example, that
objects more dense than water sink in water is a relation which is
believed to hold for all objects, past, present, and future, and not
just for those few objects that have been placed in water. It is not
possible to show that every instance of the subject class has this
property.

Aristotle's third criterion identifies causal relation and the
"essential" attribution of a predicate to a subject. This pushes back

the problem one stage. Unfortunately, Aristotle failed to provide a criterion to determine which attributions are "essential". To be sure, he did suggest that 'animal' is an essential predicate of 'man', and 'musical' is not, and that slitting an animal's throat is essentially related to its death, whereas taking a stroll is not essentially related to the occurrence of lightning.[4] But it is one thing to give examples of essential predication and accidental predication, and another thing to stipulate a general criterion for making the distinction.

The Structure of a Science

Although Aristotle did not specify a criterion of the "essential" attribution of a predicate to a subject class, he did insist that each particular science has a distinctive subject genus and set of predicates. The subject genus of physics, for example, is the class of cases in which bodies change their locations in space. Among the predicates which are proper to this science are 'position', 'speed', and 'resistance'. Aristotle emphasized that a satisfactory explanation of a phenomenon must utilize the predicates of that science to which the phenomenon belongs. It would be inappropriate, for instance, to explain the motion of a projectile in terms of such distinctively biological predicates as 'growth' and 'development'.

Aristotle held that an individual science is a deductively organized group of statements. At the highest level of generality are the first principles of *all* demonstration—the Principles of Identity, Non-Contradiction, and the Excluded Middle. These are principles applicable to *all* deductive arguments. At the next highest level of generality are the first principles and definitions of the particular science. The first principles of physics, for example, would include:

All motion is either natural or violent.
All natural motion is motion towards a natural place.
   e.g. solid objects move by nature towards the centre of the earth.
Violent motion is caused by the continuing action of an agent.
   (Action-at-a-distance is impossible.)
A vacuum is impossible.

The first principles of a science are not subject to deduction from more basic principles. They are the most general true statements that can be made about the predicates proper to the science. As such, the first principles are the starting-points of all demonstration within the science. They function as premisses for the deduction of those correlations which are found at lower levels of generality.

The Four Causes

Aristotle did place one additional requirement on scientific interpretations. He demanded that an adequate explanation of a correlation or process should specify all four aspects of causation. The four aspects are the formal cause, the material cause, the efficient cause, and the final cause.

A process susceptible to this kind of analysis is the skin-colour change of a chameleon as it moves from a bright-green leaf to a dull-grey twig. The formal cause is the pattern of the process. To describe the formal cause is to specify a generalization about the conditions under which this kind of colour change takes place. The material cause is that substance in the skin which undergoes a change of colour. The efficient cause is the transition from leaf to twig, a transition accompanied by a change in reflected light and a corresponding chemical change in the skin of the chameleon. The final cause of the process is that the chameleon should escape detection by its predators.

Aristotle insisted that every scientific explanation of a correlation or process should include an account of its final cause, or *telos*. Teleological explanations are explanations which use the expression 'in order that', or its equivalent. Aristotle required teleological explanations not only of the growth and development of living organisms, but also of the motions of inanimate objects. For example, he held that fire rises in order to reach its "natural place" (a spherical shell just inside the orbit of the moon).

Teleological interpretations need not presuppose conscious deliberation and choice. To say, for instance, that 'chameleons change colour in order to escape detection' is not to claim a conscious activity on the part of chameleons. Nor is it to claim that the behaviour of chameleons implements some "cosmic purpose".

However, teleological interpretations do presuppose that a future state of affairs determines the way in which a present state of affairs unfolds. An acorn develops in the way it does in order that it should realize its natural end as an oak-tree; a stone falls in order that it should achieve its natural end—a state of rest as near as possible to the centre of the earth; and so on. In each case, the future state "pulls along", as it were, the succession of states which leads up to it.

Aristotle criticized philosophers who sought to explain change exclusively in terms of material causes and efficient causes. He was particularly critical of the atomism of Democritus and Leucippus, in which natural processes were "explained" by the aggregation and scattering of invisible atoms. To a great extent, Aristotle's criticism was based on the atomists' neglect of final causes.

Aristotle also criticized those Pythagorean natural philosophers who believed that they had explained a process when they had found a mathematical relationship exemplified in it. According to Aristotle, the Pythagorean approach suffers from exclusive preoccupation with formal causes.

It should be added, however, that Aristotle did recognize the importance of numerical relations and geometrical relations within the science of physics. Indeed, he singled out a group of "composite sciences"—astronomy, optics, harmonics, and mechanics*—whose subject-matter is mathematical relationships among physical objects.

The Demarcation of Empirical Science

Aristotle sought, not only to mark off the subject-matter of each individual science, but also to distinguish empirical science, as a whole, from pure mathematics. He achieved this demarcation by distinguishing between applied mathematics, as practised in the composite sciences, and pure mathematics, which deals with number and figure in the abstract.

Aristotle maintained that, whereas the subject-matter of empirical science is change, the subject-matter of pure mathematics is that which is unchanging. The pure mathematician abstracts from physical situations certain quantitative aspects of bodies and their relations, and deals exclusively with these aspects. Aristotle held that these mathematical forms have no objective existence. Only in the mind of the mathematician do the forms survive the destruction of the bodies from which they are abstracted.

---

* Aristotle included mechanics in the set of composite sciences at *Posterior Analytics* 76^{a}23–5 and *Metaphysics* 1078^{a}14–7, but did not mention mechanics at *Physics* 194^{a}7–11.

The Necessary Status of First Principles

Aristotle claimed that genuine scientific knowledge has the status of necessary truth. He maintained that the properly formulated first principles of the sciences, and their deductive consequences, could not be other than true. Since first principles predicate attributes of class terms, Aristotle would seem to be committed to the following theses:

1. Certain properties inhere essentially in the individuals of certain classes; an individual would not be a member of one of these classes if it did not possess the properties in question.
2. An identity of structure exists in such cases between the universal affirmative statement which predicates an attribute of a class term, and the non-verbal inherence of the corresponding property in members of the class.
3. It is possible for the scientist to intuit correctly this isomorphism of language and reality.

Aristotle's position is plausible. We do believe that 'all men are mammals', for instance, is necessarily true, whereas 'all ravens are black' is only accidentally true. Aristotle would say that although a man could not possibly be a non-mammal, a raven might well be non-black. But, as noted above, although Aristotle did give examples of this kind to contrast "essential predication" and "accidental predication", he failed to formulate a general criterion to determine which predications are essential.

Aristotle bequeathed to his successors a faith that, because the first principles of the sciences mirror relations in nature which could not be other than they are, these principles are incapable of being false. To be sure, he could not authenticate this faith. Despite this, Aristotle's position that scientific laws state necessary truths has been widely influential in the history of science.

Notes

[1] Aristotle, *Posterior Analytics*, 89^{b}10–20.
[2] Ibid. 78^{a}38–78^{b}3.
[3] Ibid. 71^{b}20–72^{a}5.
[4] Ibid. 73^{a}25–73^{b}15.

2

The Pythagorean Orientation

PLATO (428/7–348/7 BC) was born into a distinguished Athenian family. In early life he held political ambitions, but became disillusioned, first with the tyranny of the Thirty, and then with the restored democracy which executed his friend Socrates in 399 BC. In later life, Plato made two visits to Syracuse in the hope of educating to responsible statesmanship its youthful ruler. The visits were not a success.

Plato founded the Academy in 387 BC. Under his leadership, this Athenian institution became a centre for research in mathematics, science, and political theory. Plato himself contributed dialogues that deal with the entire range of human experience. In the *Timaeus*, he presented as a "likely story" a picture of a universe structured by geometrical harmonies.

PTOLEMY (CLAUDIUS PTOLEMAEUS, *c.*100–*c.*178) was an Alexandrian astronomer about whose life virtually nothing is known. His principal work, *The Almagest*, is an encyclopedic synthesis of the results of Greek astronomy, a synthesis brought up to date with new observations. In addition, he introduced the concept of circular motion with uniform angular velocity about an equant point, a point at some distance from the centre of the circle. By using equants, in addition to epicycles and deferents, he was able to predict with fair accuracy the motions of the planets against the zodiac.

The Pythagorean View of Nature

It probably is not possible for a scientist to interrogate nature from a wholly disinterested standpoint. Even if he has no particular axe to grind, he is likely to have a distinctive way of viewing nature. The "Pythagorean Orientation" is a way of viewing nature which has been very influential in the history of science. A scientist who has this orientation believes that the "real" is the mathematical harmony that is present in nature. The committed Pythagorean is convinced that knowledge of this mathematical harmony is insight into the fundamental structure of the universe. A persuasive expression of this point of view is Galileo's declaration that

philosophy is written in this grand book—I mean the universe—which stands continually open to our gaze, but it cannot be understood unless one first learns to comprehend the language and interpret the characters in which it is written. It is written in the language of mathematics, and its characters are triangles, circles, and other geometrical figures, without which it is humanly impossible to understand a single word of it.[1]

This orientation originated in the sixth century BC when Pythagoras, or his followers, discovered that musical harmonies could be correlated with mathematical ratios, i.e.,

interval	ratio
octave	2:1
fifth	3:2
fourth	4:3

The early Pythagoreans found, moreover, that these ratios hold regardless of whether the notes are produced by vibrating strings or resonating air columns. Subsequently, Pythagorean natural philosophers read musical harmonies into the universe at large. They associated the motions of the heavenly bodies with sounds in such a way that there results a "harmony of the spheres".

Plato and the Pythagorean Orientation

Plato sometimes has been condemned for supposedly promulgating a philosophical orientation detrimental to the progress of science. The orientation in question is a turning away from the study of the world as revealed in sense experience, in favour of the contemplation of abstract ideas. Detractors of Plato often emphasize *Republic* 529–30, where Socrates recommends a shift in attention from the transient phenomena of the heavens to the timeless purity of geometrical relations. But, as Dicks has pointed out, Socrates' advice is given in the context of a discussion of the ideal education of prospective rulers. In this context, Plato is concerned to emphasize those types of study which promote the development of the capacity for abstract thought.[2] Thus he contrasts "pure geometry" with its practical application, and geometrical astronomy with the observation of light streaks in the sky.

Everyone is in agreement that Plato was dissatisfied with a "merely empirical" knowledge of the succession and coexistence of phenomena. This sort of "knowledge" must be transcended in such a way that the underlying rational order becomes manifest. The point of division among interpreters of Plato is whether it is required of the seeker of this deeper truth to turn away from what is given in sense experience. My own view is that Plato would say 'no' at this point, and would maintain that this "deeper knowledge" is to be achieved by uncovering the pattern which "lies hidden within" phenomena. At any rate, it is doubtful that Plato would have been an influence in the history of science had he not been interpreted in this manner by subsequent natural philosophers.

This influence has been expressed primarily in terms of general attitudes towards science. Natural philosophers who counted themselves "Platonists" believed in the underlying rationality of the universe and the importance of discovering it. And they drew sustenance from what they took to be Plato's similar conviction. In the late Middle Ages and the Renaissance, this Platonism was an important corrective both to the denigration of science within religious circles and to the preoccupation with disputation based on standard texts within academic circles.

In addition, commitment to Plato's philosophy tended to reinforce a Pythagorean orientation toward science. Indeed, the Pythagorean orientation became influential in the Christian West largely as a result of a marriage of Plato's *Timaeus* and Holy Scripture. In the *Timaeus*, Plato described the creation of the universe by a benevolent Demiurge, who impressed a mathematical pattern upon a formless primordial matter. This account was appropriated by Christian apologists, who identified the pattern with the Divine Plan of

Creation and repressed the emphasis on a primordial matter. For those who accepted this synthesis, the task of the natural philosopher is to uncover the mathematical pattern upon which the universe is ordered.

Plato himself suggested in the *Timaeus* that the five "elements"— four terrestrial and one celestial—may be correlated with the five regular solids.

| Tetrahedron | Cube | Octahedron | Icosahedron | Dodecahedron |
| (fire) | (earth) | (air) | (water) | (celestial matter) |

He assigned the tetrahedron to fire, because the tetrahedron is the regular solid with the sharpest angles, and because fire is the most penetrating of elements. He assigned the cube to Earth, because it takes more effort to tip over a cube on its base than it does to tip over any one of the remaining three regular solids, and because Earth is the most "solid" of the elements. Plato used similar reasoning to assign the octahedron to air, the icosahedron to water, and the dodecahedron to celestial matter. In addition, he suggested that transformations among water, air, and fire result from a "dissolution" of each equilateral triangular face of the respective regular solids into six 30−60−90-degree triangles,* with subsequent recombination of these smaller triangles to form the faces of other regular solids. Plato's explanation of matter and its properties in terms of geometrical figures is very much in the Pythagorean tradition.

The Tradition of "Saving the Appearances"

The Pythagorean natural philosopher believes that mathematical relations which fit phenomena count as explanations of why things

* Viz.:

are as they are. This point of view has had opposition, almost from its inception, from a rival point of view. This rival view is that mathematical hypotheses must be distinguished from theories about the structure of the universe. On this view, it is one thing to "save the appearances" by superimposing mathematical relations on phenomena, but quite another thing to explain why the phenomena are as they are.

This distinction between physically true theories and hypotheses which save the appearances was made by Geminus in the first century BC. Geminus outlined two approaches to the study of celestial phenomena. One is the approach of the physicist, who derives the motions of the heavenly bodies from their essential natures. The second is the approach of the astronomer, who derives the motions of the heavenly bodies from mathematical figures and motions. He declared that

it is no part of the business of an astronomer to know what is by nature suited to a position of rest, and what sort of bodies are apt to move, but he introduces hypotheses under which some bodies remain fixed, while others move, and then considers to which hypotheses the phenomena actually observed in the heaven will correspond.[3]

Ptolemy on Mathematical Models

In the second century AD, Claudius Ptolemy formulated a series of mathematical models, one for each planet then known. One important feature of the models is the use of epicycle-deferent circles to reproduce the apparent motions of the planets against the zodiac. On the epicycle-deferent model, the planet P moves along an epicyclic circle, the centre of which moves along a deferent circle around the earth. By adjusting the speeds of revolution of the points P and C, Ptolemy could reproduce the observed periodic retrograde motion of the planet. In passing from A to B along the epicycle, the planet appears to an observer on earth to reverse the direction of its motion against the background stars.

Ptolemy emphasized that more than one mathematical model can be constructed to save the appearances of planetary motions. He noted, in particular, that a moving-eccentric system can be constructed which is mathematically equivalent to a given epicycle-deferent system.*

* Ptolemy credited Apollonius of Perga (*fl.* 220 BC) with the first demonstration of this equivalence.

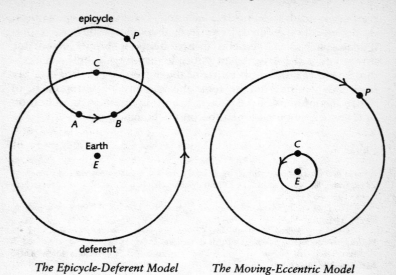

The Epicycle-Deferent Model   The Moving-Eccentric Model

In the moving-eccentric model, planet *P* moves along a circle centred on eccentric point *C*, which point *C* moves, with opposite-directed motion, along a circle centred on the Earth *E*. Since the two models are mathematically equivalent, the astronomer is at liberty to employ whichever model is the more convenient.

A tradition arose in astronomy that the astronomer should construct mathematical models to save the appearances, but should not theorize about the "real motions" of the planets. This tradition owed much to Ptolemy's work on planetary motions. Ptolemy himself, however, did not consistently defend this position. He did hint in the *Almagest* that his mathematical models were computational devices only, and that he was not to be understood as claiming that the planets actually describe epicyclic motions in physical space. But in a later work, the *Hypotheses Planetarum*, he claimed that his complicated system of circles revealed the structure of physical reality.

Ptolemy's uneasiness about restricting astronomy to saving the appearances was echoed by Proclus, a fifth-century Neoplatonist. Proclus complained that astronomers had subverted proper scientific method. Instead of deducing conclusions from self-evident axioms, on the model of geometry, they frame hypotheses solely to accommodate phenomena. Proclus insisted that the proper axiom for astronomy is the Aristotelian principle that every simple motion is motion either around the centre of the universe or toward or away from this centre. And he took the inability of astronomers to derive the motions of the planets from this axiom as an indication of a divinely imposed limitation on the human mind.

Notes

[1] Galileo, *The Assayer*, trans. by S. Drake, in *The Controversy on the Comets of 1618*, trans. S. Drake and C. D. O'Malley (Philadelphia: University of Pennsylvania Press, 1960), 183–4.

[2] D. R. Dicks, *Early Greek Astronomy to Aristotle* (London: Thames and Hudson, 1970), 104–7.

[3] Geminus is quoted by Simplicius, *Commentary on Aristotle's Physics*, in T. L. Heath, *Aristarchus of Samos* (Oxford: Clarendon Press, 1913), 275–6; reprinted in *A Source Book in Greek Science*, ed. M. Cohen and I. E. Drabkin (New York: McGraw-Hill, 1948), 91.

3

The Ideal of Deductive Systematization

EUCLID (*fl.* 300 BC), according to Proclus, taught and founded a school at Alexandria. His most important surviving work is the *Elements*. It is not possible to say with any assurance to what extent this work was a codification of existing geometrical knowledge and to what extent it was the fruit of original research. It seems likely that, in addition to setting out geometry as a deductive system, Euclid constructed a number of original proofs.

ARCHIMEDES (287–212 BC), the son of an astronomer, was born at Syracuse. It is believed that he spent some time at Alexandria, perhaps studying with the successors of Euclid. Upon his return to Syracuse, he devoted himself to research in pure and applied mathematics.

Archimedes' fame in Antiquity derived in large measure from his prowess as a military engineer. It is reported that catapults of his design were used effectively against the Romans during the siege of Syracuse. Archimedes himself was said to prize more highly his abstract investigations of conic sections, hydrostatics, and equilibria involving the law of the lever. According to legend, Archimedes was slain by Roman soldiers while he was contemplating a geometrical problem.

A widely held thesis among ancient writers was that the structure of a completed science ought to be a deductive system of statements. Aristotle had emphasized the deduction of conclusions from first principles. Many writers in late Antiquity believed that the ideal of deductive systematization had been realized in the geometry of Euclid and the statics of Archimedes.

Euclid and Archimedes had formulated systems of statements— comprising axioms, definitions, and theorems—organized so that the truth of the theorems follows from the assumed truth of the axioms. For example, Euclid proved that his axioms, together with definitions of such terms as 'angle' and 'triangle', imply that the sum of the angles of a triangle is equal to two right angles. And Archimedes proved from his axioms on the lever that two unequal weights balance at distances from the fulcrum that are inversely proportional to their weights.

Three aspects of the ideal of deductive systematization are (1) that the axioms and theorems are deductively related; (2) that the axioms themselves are self-evident truths; and (3) that the theorems agree with observations. Philosophers of science have taken different positions on the second and third aspects, but there has been general agreement on the first aspect.

One cannot subscribe to the deductive ideal without accepting the requirement that theorems be related deductively to axioms. Euclid and Archimedes utilized two important techniques to prove theorems from their axioms: *reductio ad absurdum* arguments, and a method of exhaustion.

The *reductio ad absurdum* technique of proving theorem '*T*' is to assume that 'not *T*' is true and then deduce from 'not *T*' and the axioms of the system both a statement and its negation. If two contradictory statements can be deduced in this way, and if the axioms of the system are true, then '*T*' must be true as well.*

The method of exhaustion is an extension of the *reductio ad absurdum* technique. It consists of showing that each possible contrary of a theorem has consequences that are inconsistent with the axioms of a system.†

* Archimedes used a *reductio ad absurdum* argument to prove that 'weights that balance at equal distances from a fulcrum are equal' ('*T*'). He began by assuming the truth of the contradictory statement that 'the balancing weights are of unequal magnitude' ('not *T*'), and then showed that 'not *T*' is false, because it has implications that contradict one of the axioms of the system. For if 'not *T*' were true, one could decrease the weight of the greater so that the two weights were of equal magnitude. But axiom 3 states that, if one of two weights initially in equilibrium is decreased, then the lever inclines toward the undiminished weight. The lever no longer would be in equilibrium. But this contradicts 'not *T*', thereby establishing '*T*'.[1]

† Archimedes used the method of exhaustion to prove that the area of a circle is equal to the area of a right triangle whose base is the radius of the circle and whose altitude is its circumference. Archimedes proved this theorem by showing that, if one assumes that the area of the circle either is greater than or is less than that of the triangle, contradictions ensue within the axiom system of geometry.[2] See diagram on page 25.

With regard to the requirement of deductive relations between axioms and theorems, Euclid's geometry was deficient. Euclid deduced a number of his theorems by appealing to the operation of superimposing figures to establish their congruence. But no reference is made in the axioms to this operation of superposition. Thus Euclid "proved" some of his theorems by going outside the axiom system. Euclid's geometry was recast into rigorous deductive form by David Hilbert in the latter part of the nineteenth century. In Hilbert's reformulation, every theorem of the system is a deductive consequence of the axioms and definitions.

A second, more controversial aspect of the ideal of deductive systematization is the requirement that the axioms themselves be self-evident truths. This requirement was stated clearly by Aristotle, who insisted that the first principles of the respective sciences be necessary truths.

The requirement that the axioms of deductive systems be self-evident truths was consistent with the Pythagorean approach to natural philosophy as well. The committed Pythagorean believes that there exist in nature mathematical relations that can be discovered by reason. From this standpoint, it is natural to insist that the starting-points of deductive systematization be those mathematical relations which have been found to underlie phenomena.

A different attitude was taken by those who followed the tradition of saving the appearances in mathematical astronomy. They rejected the Aristotelian requirement. To save appearances it suffices that

Archimedes' Circle–Triangle Relation

the deductive consequences of the axioms should agree with observations. That the axioms themselves are implausible, or even false, is irrelevant.

The third aspect of the ideal of deductive systematization is that the deductive system should make contact with reality. Certainly Euclid and Archimedes intended to prove theorems which had practical application. Indeed Archimedes was famous for his application of the law of the lever to the construction of catapults for military purposes.

But to make contact with the realm of experience it is necessary that at least some of the terms of the deductive system should refer to objects and relations in the world. It seems just to have been assumed by Euclid, Archimedes, and their immediate successors that such terms as 'point', 'line', 'weight', and 'rod' do have empirical correlates. Archimedes, for instance, does not mention the problems involved in giving an empirical interpretation to his theorems on the lever. He made no comments on the limitations that must be imposed on the nature of the lever itself. And yet the theorems he derived are confirmed experimentally only for rods that do not bend appreciably, and which have a uniform weight distribution. Archimedes' theorems apply strictly only to an "idealized lever" which, in principle, cannot be realized in experience, namely, an infinitely rigid, but mass-less, rod.

It may be that Archimedes' preoccupation with laws applicable to this "ideal lever" reflects a philosophical tradition in which a contrast is drawn between the unruly complexities of phenomena and the timeless purity of formal relationships. This tradition often was reinforced by the ontological claim that the phenomenal realm is at best an "imitation" or "reflection" of the "real world". Primary responsibility for promulgating this point of view rests with Plato and his interpreters. This dualism had important repercussions in the thought of Galileo and Descartes.

Notes

[1] T. L. Heath (ed.), *The Works of Archimedes* (New York: Dover Publications, 1912), 189–90.
[2] Ibid. 91–3.

4

Atomism and the Concept of Underlying Mechanism

As noted above, some followers of Plato construed the world to be an imperfect reflection of an underlying reality. A more radical discontinuity was suggested by the atomists Democritus and Leucippus. For the atomists, the relation between appearance and reality was not the relation between an original and an imperfect copy. Rather, they believed that objects and relations in the "real world" were *different in kind* from the world we know by means of the senses.

What is real, according to the atomists, is the motion of atoms through the void. It is the motions of atoms which cause our perceptual experience of colours, odours, and tastes. Were there no such motions, there would be no perceptual experience. Moreover, the atoms themselves have only the properties of size, shape, impenetrability, and motion, and the propensity to enter into various combinations and associations. Unlike macroscopic objects, atoms can be neither penetrated nor subdivided.

The atomists attributed phenomenal changes to the association and dissociation of atoms. For instance, they attributed the salty taste of some foodstuffs to the setting free of large, jagged atoms, and the ability of fire to penetrate bodies to the rapid motions of tiny, spherical fire-atoms.[1]

Several aspects of the atomists' programme have been important in the development of subsequent views of scientific method. One influential aspect of atomism is the idea that observed changes can be explained by reference to processes occurring at a more elementary level of organization. This became an item of belief for many natural philosophers in the seventeenth century. That submacroscopic interactions cause macroscopic changes was affirmed by Gassendi, Boyle, and Newton, among others.

Moreover, the ancient atomists realized, tacitly at least, that one cannot explain adequately qualities and processes at one level merely by postulating that the same qualities and processes are present at a deeper level. For instance, one cannot account satisfactorily for the colours of objects by attributing the colours to the presence of coloured atoms.

A further important aspect of the atomists' programme is the reduction of qualitative changes at the macroscopic level to quantitative changes at the atomic level. Atomists agreed with Pythagoreans that scientific explanations ought to be given in terms of geometrical and numerical relationships.

Two factors weighed against any widespread acceptance of the classical version of atomism. The first factor was the uncompromising materialism of this philosophy. By explaining sensation and even thought in terms of the motions of atoms, the atomists challenged man's self-understanding. Atomism seemed to leave no place for spiritual values. Surely the values of friendship, courage, and worship cannot be reduced to the concourse of atoms. Moreover, the atomists left no place in science for considerations of purpose, whether natural or divine.

The second factor was the *ad hoc* nature of the atomists' explanations. They offered a picture-preference, a way of looking at phenomena, but there was no way to check the accuracy of the picture. Consider the dissolving of salt in water. The strongest argument advanced by classical atomists was that the effect *could be* produced by dispersal of salt-atoms into the liquid. However, the classical atomists could not explain why salt dissolves in water whereas sand does not. Of course they could say that salt-atoms fit into the interstices between water-atoms whereas sand-atoms do not. But the critics of atomism would dismiss this "explanation" as merely another way of saying that salt dissolves in water whereas sand does not.

Note

[1] G. S. Kirk and J. E. Raven, *The Presocratic Philosophers* (Cambridge: Cambridge University Press, 1962), 420–3.

5

Affirmation and Development of Aristotle's Method in the Medieval Period

ROBERT GROSSETESTE (c.1168–1253) was a scholar and teacher at Oxford who became a statesman of the Church. He was Chancellor of Oxford University (1215–21), and from 1224 served as lecturer in philosophy to the Franciscan order. Grosseteste was the first medieval scholar to analyse the problems of induction and verification. He wrote commentaries on Aristotle's *Posterior Analytics and Physics*, prepared translations of *De Caelo* and *Nicomachean Ethics*, and composed treatises on calendar reform, optics, heat, and sound. He developed a Neoplatonic "metaphysics of light" in which causal agency is attributed to the multiplication and outward spherical diffusion of "species", upon analogy to the propagation of light from a source. Grosseteste became Bishop of Lincoln in 1235 and redirected his considerable energies so as to include ecclesiastical administration.

ROGER BACON (*c.*1214–92) studied at Oxford and then Paris, where he taught and wrote analyses of various Aristotelian works. In 1247 he returned to Oxford, where he studied various languages and the sciences, with particular emphasis on optics. Pope Clement IV, on learning of Bacon's proposed unification of the sciences in the service of theology, requested a copy of Bacon's work. Bacon had not yet put his views on paper, but he rapidly composed and dispatched to the Pope the *Opus Maius* and two companion works (1268). Unfortunately the Pope died before having assessed Bacon's contribution.

Bacon appears to have antagonized his superiors in the Franciscan order by his sharp criticism of the intellectual capabilities of his colleagues. Moreover, his enthusiasm for alchemy, astrology, and the apocalypticism of Joachim of Floris rendered him suspect. It is likely, although not beyond doubt, that he spent several of his later years under confinement.

JOHN DUNS SCOTUS (*c.*1265–1308) entered the Franciscan order in 1280 and was ordained a priest in 1291. He studied at Oxford and Paris, where he received a doctorate in theology in 1305, despite having been banished from Paris for a time for failing to support the King in a dispute with the Pope over the taxation of Church lands. In company with many other medieval writers, Duns Scotus sought to assimilate Aristotelian philosophy to Christian doctrine.

WILLIAM OF OCKHAM (*c.*1280–1349) studied and taught at Oxford. He soon became a focus of controversy within the Church. He attacked the Pope's claim of temporal supremacy, insisting on the divinely ordained independence of civil authority. He appealed to the prior pronouncements of Pope Nicholas III in a dispute with Pope John XXII over apostolic poverty. And he defended the nominalist position that universals have objective value only in so far as they are present in the mind. Ockham took refuge in Bavaria for a time while his writings were under examination at Avignon. No formal condemnation took place, however.

NICOLAUS OF AUTRECOURT (*c.*1300–after 1350) studied and lectured at the University of Paris, where he developed a critique of the prevalent doctrines of substance and causality. In 1346 he was sentenced by the

Avignon Curia to burn his writings and to recant certain condemned doctrines before the faculty of the University of Paris. Nicolaus complied, and, curiously enough, subsequently was appointed deacon at the Cathedral of Metz (1350).

Prior to 1150, Aristotle was known to scholars in the Latin West primarily as a logician. Plato was held to be the pre-eminent philosopher of nature. But commencing about 1150, Aristotle's writings on science and scientific method began to be translated from Arabic and Greek sources into Latin. Centres of translating activity arose in Spain and Italy. By 1270, the extensive Aristotelian corpus had been translated into Latin. The impact of this achievement on intellectual life in the West was very great indeed. Aristotle's writings on science and scientific method provided scholars with a wealth of new insights. So much so that for several generations the standard presentation of a work on a particular science took the form of a commentary on the corresponding study by Aristotle.

Aristotle's most important writing on the philosophy of science is the *Posterior Analytics*, a work that became available to western scholars in the latter part of the twelfth century. During the next three centuries, writers on scientific method addressed themselves to the problems that had been formulated by Aristotle. In particular, medieval commentators discussed and criticized Aristotle's view of scientific procedure, his position on evaluating competing explanations, and his claim that scientific knowledge is necessary truth.

The Inductive–Deductive Pattern of Scientific Inquiry

Robert Grosseteste and Roger Bacon, the two most influential thirteenth-century writers on scientific method, affirmed Aristotle's inductive–deductive pattern of scientific inquiry. Grosseteste referred to the inductive stage as a "resolution" of phenomena into constituent elements, and to the deductive stage as a "composition" in which these elements are combined to reconstruct the original phenomena.[1] Subsequent writers often referred to Aristotle's theory of scientific procedure as the "Method of Resolution and Composition".

Grosseteste applied the Aristotelian theory of procedure to the problem of spectral colours. He noted that the spectra seen in rainbows, mill-wheel sprays, boat-oar sprays, and the spectra produced by passing sunlight through water-filled glass spheres, shared certain common characteristics. Proceeding by induction, he "resolved" three elements which are common to the various instances. These elements are (1) that the spectra are associated with transparent spheres, (2) that different colours result from the refraction of light through different angles, and (3) that the colours produced lie on the arc of a circle. He then was able to "compose" the general features of this class of phenomena from the above three elements.[2]

Roger Bacon's "Second Prerogative" of Experimental Science

Grosseteste's Method of Resolution specifies an inductive ascent from statements about phenomena to elements from which the phenomena may be reconstructed. Grosseteste's pupil Roger Bacon emphasized that successful application of this inductive procedure depends on accurate and extensive factual knowledge. Bacon suggested that the factual base of a science often may be augmented by active experimentation. The use of experimentation to increase knowledge of phenomena is the second of Bacon's "Three Prerogatives of Experimental Science".[3]

Bacon praised a certain "master of experimentation" whose work constituted a realization of the second prerogative. The individual cited probably was Petrus of Maricourt.[4] Petrus had demonstrated, among other things, that breaking a magnetic needle crosswise into two fragments produces two new magnets, each with its own north pole and south pole. Bacon emphasized that discoveries such as this increase the observational base from which the elements of magnetism may be induced.

Had Bacon restricted his praise of experimentation to this kind of investigation, he would merit recognition as a champion of experimental inquiry. However, Bacon often placed experimentation in the service of alchemy, and he made extravagant and unsupported claims for the results of alchemical experiments. He declared, for instance, that one triumph of "Experimental Science" was the discovery of a substance that removes the impurities from base metals such that pure gold remains.[5]

The Inductive Methods of Agreement and Difference

Aristotle had insisted that explanatory principles should be induced from observations. An important contribution of medieval scholars was to outline additional inductive techniques for discovering explanatory principles.

Robert Grosseteste, for example, suggested that one good way to determine whether a particular herb has a purgative effect would be to examine numerous cases in which the herb is administered under conditions where no other purgative agents are present.[6] It would be difficult to implement this test, and there is no evidence that Grosseteste attempted to do so. But he must be credited with outlining an inductive procedure which centuries later came to be known as "Mill's Joint Method of Agreement and Difference".

In the fourteenth century, John Duns Scotus outlined an inductive Method of Agreement, and William of Ockham outlined an inductive Method of Difference. They regarded these methods as aids in the "resolution" of phenomena. As such, they are procedures intended to supplement the inductive procedures which Aristotle had discussed.

Duns Scotus's Method of Agreement

Duns Scotus's Method of Agreement is a technique for analysing a number of instances in which a particular effect occurs. The procedure is to list the various circumstances that are present each time the effect occurs, and to look for some one circumstance that is present in every instance.[7] Duns Scotus would hold that, if a listing of circumstances has the form

Instance	Circumstances	Effect
1	*ABCD*	*e*
2	*ACE*	*e*
3	*ABEF*	*e*
4	*ADF*	*e*

then the investigator is entitled to conclude that *e can be* the effect of cause *A*.

Duns Scotus's claims for his Method of Agreement were quite modest. He held that the most that can be established by an application of the method is an "aptitudinal union" between an effect and an accompanying circumstance. By applying the schema, a scientist may conclude, for instance, that the moon is a body that *can be* eclipsed, or that a certain kind of herb *can have* a bitter taste.[8] But application of the schema alone can establish neither that the moon necessarily must be eclipsed, nor that every sample of the herb necessarily is bitter.

Paradoxically, Duns Scotus both augmented the Method of Resolution and undercut confidence in inductively established correlations. His theological convictions were responsible for the latter emphasis. He insisted that God can accomplish anything which does not involve a contradiction, and that uniformities in nature exist only by the forbearance of God. Moreover, God could, if He wished, short-circuit a regularity and produce an effect directly without the presence of the usual cause. It was for this reason that Duns Scotus held that the Method of Agreement can establish only aptitudinal unions within experience.

William of Ockham's Method of Difference

Emphasis on the omnipotence of God is still more pronounced in the writings of William of Ockham. Ockham repeatedly insisted that God can accomplish anything that can be done without contradiction. In agreement with Duns Scotus, he held that the scientist can establish by induction only aptitudinal unions among phenomena.

Ockham formulated a procedure for drawing conclusions about aptitudinal unions according to a Method of Difference. Ockham's method is to compare two instances—one instance in which the effect is present, and a second instance in which the effect is not present. If it can be shown that there is a circumstance present when the effect is present and absent when the effect is absent, e.g.,

Instance	Circumstances	Effect
1	ABC	e
2	AB	—

then the investigator is entitled to conclude that the circumstance C *can be* the cause of effect *e*.

Ockham maintained that, in the ideal case, knowledge of an aptitudinal union can be established on the basis of just one observed association. He noted, though, that in such a case one would have to be certain that all other possible causes of the effect in question are absent. He observed that in practice it is difficult to determine whether two sets of circumstances differ in one respect only. For that reason, he urged that numerous cases be investigated in order to minimize the possibility that some unrecognized factor is responsible for the occurrence of the effect.[9]

Evaluation of Competing Explanations

Grosseteste and Roger Bacon, in addition to restating Aristotle's inductive–deductive pattern of scientific inquiry, also made original contributions to the problem of evaluating competing explanations. They recognized that a statement about an effect may be deduced from more than one set of premises. Aristotle, too, had been aware of this, and had insisted that genuine scientific explanations state causal relationships.

Roger Bacon's "First Prerogative" of Experimental Science

Both Grosseteste and Bacon recommended that a third stage of inquiry be added to Aristotle's inductive–deductive procedure. In this third stage of inquiry, the principles induced by "resolution" are submitted to the test of further experience. Bacon called this testing procedure the "first prerogative" of experimental science.[10] This was a valuable methodological insight, and constituted a significant advance over Aristotle's theory of procedure. Aristotle had been content to deduce statements about the same phenomena which serve as the starting-points of an investigation. Grosseteste and Bacon demanded further experimental testing of the principles reached by induction.

At the beginning of the fourteenth century, Theodoric of Freiberg made a striking application of Bacon's first prerogative. Theodoric believed that the rainbow is caused by a combination of refraction and reflection of sunlight by individual raindrops. In order to test this hypothesis, he filled hollow crystalline spheres with water, and placed them in the path of the sun's rays. He reproduced with these model drops both primary rainbows and secondary rainbows. Theodoric demonstrated that the reproduced secondary rainbows had their order of colours reversed, and that the angle between incident and emergent rays for the secondary rainbows was eleven degrees greater than for the primary rainbows. This is in good agreement with what is observed in naturally occurring rainbows.[11]

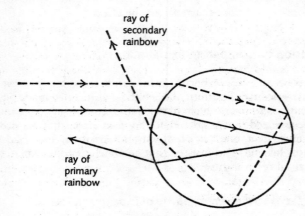

Theodoric's Model Raindrop

Unfortunately, Grosseteste and Bacon themselves frequently ignored their own counsel. Bacon, in particular, often appealed to *a priori* considerations and the authority of previous writers, rather than to additional experimental testing. For example, after declaring that experimental science is admirably suited to establish conclusions about the nature of the rainbow, Bacon insisted that there must be just five colours in the rainbow because the number five is the ideal number to set forth a variation in qualities.[12]

Grosseteste's Method of Falsification

Grosseteste noted that if a statement about an effect can be deduced from more than one set of premises, then the best approach is to eliminate all but one of the explanations. He maintained that if a hypothesis implies certain consequences, and if these consequences can be shown to be false, then the hypothesis itself must be false. Logicians have given the name '*modus tollens*' to this type of deductive argument:

> If H then C
> not C
> _____
> $\therefore$ not H

Given a group of hypotheses, each of which can be used as a premiss to deduce a given effect, it may be possible to eliminate all hypotheses but one by means of *modus tollens* arguments. To do this one would have to show that every hypothesis but one implies *other* consequences which are known to be false.

Grosseteste applied the Method of Falsification to support a hypothesis about the generation of the sun's heat. According to Grosseteste, there are just three modes of generating heat: by conduction from a hot body, "by motion", and by a concentration of rays. He believed that the sun generates heat by a concentration of rays, and he sought to exclude the other two possibilities by *modus tollens* arguments. He "falsified" the conduction hypothesis by the following argument:

> If the sun generates heat by conduction,
>   then the adjacent celestial matter is
>   heated and undergoes a change of quality.
> But the adjacent celestial matter is
>   immutable and does not undergo a change
>   of quality.
> _____
> Therefore, the sun does not generate
>   heat by conduction.[13]

This argument has the *modus tollens* form, and hence is valid—if its premisses are true, then its conclusion must be true as well. However, the second premiss, which asserts the immutability of the adjacent celestial matter, is false. Grosseteste's argument did not prove false the conduction hypothesis. And his argument to falsify the motion hypothesis failed for a similar reason.[14]

Grosseteste was not the first scholar to use *modus tollens* arguments to falsify rival hypotheses. Philosophers and math-

ematicians had made use of this technique since the time of Euclid.*
Grosseteste's achievement was the systematic application of this
technique to supplement Aristotle's evaluation procedures for
scientific hypotheses.

In spite of the fact that Grosseteste's numerous applications of
modus tollens arguments are unconvincing in the light of current
scientific knowledge, the method of falsification itself was widely
influential. The fourteenth-century scholar John Buridan, for
example, used a *modus tollens* argument to falsify a hypothesis
about projectile motion that had been mentioned, but not defended,
by Aristotle. On this hypothesis, the air in front of the projected
body rushes around to the rear in order to prevent the occurrence
of a vacuum, thereby pushing forward the projectile. Buridan
pointed out that if this hypothesis were true, then a projectile
with a blunt posterior end should move faster than one with two
pointed ends. He insisted that a projectile with a blunt posterior
end does not travel faster, although he did not claim to have
performed experiments with the two types of projectiles.[16]

Ockham's "Razor"

A large number of medieval writers defended the principle that
nature always chooses the simplest path. Grosseteste, for instance,
maintained that the angle of refraction must be one half of the
angle of incidence for a light ray passing into a denser medium. He
believed that this $1:2$ ratio holds because nature pursues the
simplest course, and because the $1:1$ ratio is unavailable since it
governs reflection.[17]

William of Ockham opposed this tendency to read into nature

* An example is Euclid's proof that there is no greatest prime number. Euclid
began by assuming the contradictory: that there does exist a greatest prime number,
denoted by N. He then formed the number

$$N' = (2 \times 3 \times 5 \times 7 \times 11 \times \dots N) + 1,$$

in which the product within parentheses includes every prime number up to and
including N. He then formed the following *modus tollens* argument:

If N is the greatest prime number, then N' (which is greater than N) is not a prime
number.

But N' is a prime number (since division of N' by any prime number leaves a
remainder of 1).

Therefore, N is not the greatest prime number.[15]

human ideas about simplicity. He felt that to insist that nature always follows the simplest path is to limit God's power. God may very well choose to achieve effects in the most complicated of ways.

For this reason, Ockham shifted emphasis on simplicity from the course of nature to theories which are formulated about it. Ockham used simplicity as a criterion of concept-formation and theory-construction. He held that superfluous concepts are to be eliminated, and suggested that the simpler of two theories that account for a type of phenomena is to be preferred. Subsequent writers often referred to this methodological principle as "Ockham's Razor".

Ockham applied his Razor in the medieval debates on the nature of projectile motion. One view was that a projectile's motion is caused by an acquired "impetus" which resides somehow in the projectile as long as it is in motion. Ockham held that impetus is a superfluous concept. According to Ockham, a statement about the 'motion of a body' is shorthand for a series of statements that attribute to the body various positions at various times. And motion is not a property of a body, but is a relation which a body has to other bodies and to time. Since change of position is not a "property" of a body, there is no need to assign an efficient cause to this relative displacement. Ockham maintained that to say "a body moves because of an acquired impetus" is to say no more than "a body moves", and he recommended elimination from physics of the concept of impetus.[18]

The Controversy about Necessary Truth

Aristotle had insisted that because a "natural necessity" orders the relations among the species and genera of objects and events, the appropriate verbal expression of these relations must have the status of necessary truth. According to Aristotle, the first principles of the sciences are not merely contingently true. They are incapable of being false, because they mirror relations in nature which could not be other than they are.

An important fourteenth-century development in the philosophy of science was a reassessment of the cognitive status of scientific interpretations. John Duns Scotus, William of Ockham, and

Nicolaus of Autrecourt, among others, sought to determine what kinds of statements, if any, are necessary truths. Their point of departure was Aristotle's position that the first principles of the sciences are self-evident, necessary representations of the way things are.

Duns Scotus on the "Aptitudinal Union" of Phenomena

Duns Scotus insisted on a distinction between the origin of first principles and the warrant for their status as necessary truths. He agreed with Aristotle that knowledge of first principles arises out of sense experience, but he added that the necessary status of these principles is independent of the truth of reports about sense experience. According to Duns Scotus, sense experience provides occasions for recognizing the truth of a first principle, but sense experience is not evidence for this truth. Rather, a first principle is true in virtue of the meanings of its constituent terms. This is so, despite the fact that it is from experience that we learn the meanings of these terms.[19] For instance, that 'opaque bodies cast shadows' is self-evident to anyone who understands the meanings of the terms 'opaque', 'cast', and 'shadow'. Moreover, this principle is a necessary truth. To deny it is to formulate a self-contradiction. Duns Scotus held that not even God could cause a self-contradiction to be implemented in the world.

Duns Scotus held that two types of scientific generalizations are necessary truths: the first principles and their deductive consequences, and statements of aptitudinal unions of phenomena. By contrast, he held that empirical generalizations are contingent truths. For example, it is necessarily true that all ravens *can be* black, but it is only a matter of contingent fact that all ravens examined have been black.

Of course the scientist cannot rest content with knowledge of aptitudinal unions of phenomena. To say that ravens *can be* black or that the moon *can be* eclipsed is to say relatively little about ravens and the moon. Duns Scotus recognized this. He recommended that, wherever possible, generalizations be deduced from first principles. The two examples differ in this respect. That the moon is a body frequently eclipsed may be deduced from the first principles that opaque bodies cast shadows, and that the earth is an opaque body which frequently is interposed between the luminous sun and the moon. No such derivation is available in the case of black ravens.

Nicolaus of Autrecourt on Necessary Truth as Conforming to the Principle of Non-Contradiction

Nicolaus of Autrecourt restricted the range of certain knowledge more severely than did Duns Scotus. Nicolaus's analysis was the culmination of a fourteenth-century erosion of confidence in what can be known to be necessarily true.

Nicolaus resolved to accept as necessary truths only those judgements that satisfy the Principle of Non-Contradiction. Following Aristotle, he announced that the primary principle of reasoning is that contradictories cannot both be true.

But although Aristotle did state that the Principle of Non-Contradiction is the ultimate principle of all demonstration, he also recognized that no conclusions about physical or biological phenomena can be deduced from this principle alone. Hence Aristotle included among the first principles of demonstration both general logical principles such as the Laws of Identity, Non-Contradiction, and the Excluded Middle, and first principles proper to the respective sciences.

Nicolaus, however, refused to concede certainty to the inductively established first principles of the sciences, whether these principles state causal relations or mere aptitudinal unions of phenomena. He restricted certain knowledge to the Principle of Non-Contradiction itself and those statements and arguments that "conform" to it. The only exceptions he allowed were the articles of faith.[20]

Nicolaus insisted that every scientific demonstration should conform to the principle that every statement of the form 'A and not A' is necessarily false. According to Nicolaus, an argument "conforms" to the Principle of Non-Contradiction if, and only if, the conjunction of its premises and the negation of its conclusion

$$(P_1 \cdot P_2 \cdot P_3 \cdot \ldots P_n) \cdot {\sim}C'$$

is a self-contradiction.* Logicians today accept this requirement as a necessary and sufficient condition of deductive validity.

---

* The symbol '·' stands for the English 'and' in conjunctions of the form 'p and q' where p and q are individual sentences. The expression '${\sim}p$' stands for the English 'It is false that p'.

Nicolaus held that every valid argument is reducible to the Principle of Non-Contradiction either immediately or mediately. The reduction is immediate if the conclusion is identical with the premises or a part of the premises. For example, it is immediately evident that arguments of the form $\dfrac{A}{\therefore A}$ and $\dfrac{A \cdot B \cdot C}{\therefore A}$ satisfy the Principle of Non-Contradiction. The reduction is mediate in the case of syllogistic arguments. For example, given the syllogism

P_1— All quadrilaterals are polygons.
P_2— All squares are quadrilaterals.

C —$\therefore$ All squares are polygons.

the negation of the conclusion is inconsistent with the conjunction of the premises. However, it is not immediately evident that the statement '$(P_1 \cdot P_2) \cdot \sim C$' is a self-contradiction. The statement is a self-contradiction only because '$(P_1 \cdot P_2)$' implies 'C'.

On the basis of this analysis of the nature of deductive arguments, Nicolaus denied that a necessary knowledge of causal relations could be achieved. He pointed out that no information can be deduced from a set of premises except that information implied by, or "contained in", the premises. In this respect, deductive arguments are like orange-juicers—no more juice can be extracted than is present initially in the oranges. But since a cause is something distinct from its effect, one cannot deduce a statement about an effect from statements about its supposed cause. Nicolaus insisted that it is not possible to deduce that because a particular phenomenon occurred, it must be accompanied by, or followed by, some other phenomenon.

Nicolaus argued, moreover, that it is not possible to achieve a necessary knowledge of causal relations by application of the Method of Agreement. He insisted that it cannot be established that a correlation which has been observed to hold must continue to hold in the future.[21] Duns Scotus, of course, could have accepted Nicolaus's critique without abandoning his own position, for he claimed to establish only aptitudinal unions between two types of phenomena.

The conclusion of Nicolaus's analysis is that no necessary knowledge of causal relations can be achieved. Statements about causes do not imply statements about effects, and inductive arguments do not prove that an observed correlation must hold.

Nicolaus declared that he hoped that his critique of what can be known with certainty would be of service to the Christian faith. He noted with disapproval that scholars spent entire lifetimes in the study of Aristotle. He suggested that it would be better if this energy were expended to improve the faith and morals of the community.[22] Perhaps for this reason, he appended to his critique a "probable" theory of the universe based on classical atomism. Nicolaus wished to show, not only that Aristotle's science was not a science of certainties, but also that Aristotle's view of the universe was not even the most probable of world-views.

Notes

[1] A. C. Crombie, *Robert Grosseteste and the Origins of Experimental Science (1100–1700)* (Oxford: Clarendon Press, 1953), 52–66.

[2] Ibid. 64–6.

[3] Roger Bacon, *The Opus Majus*, trans. Robert B. Burke (New York: Russell and Russell, 1962), ii. 615–16.

[4] See e.g. A. C. Crombie, *Robert Grosseteste*, 204–10.

[5] Roger Bacon, *The Opus Majus*, ii. 626–7.

[6] A. C. Crombie, *Robert Grosseteste*, 73–4.

[7] *Duns Scotus: Philosophical Writings*, trans. and ed. Allan Wolter (Edinburgh: Thomas Nelson, 1962), 109.

[8] Ibid. 110–11.

[9] See e.g. Julius R. Weinberg, *Abstraction, Relation and Induction* (Madison, Wis.: The University of Wisconsin Press, 1965), 145–7.

[10] Roger Bacon, *The Opus Majus*, ii. 587.

[11] See A. C. Crombie, *Robert Grosseteste*, 233–59; W. A. Wallace, *The Scientific Methodology of Theodoric of Freiberg* (Fribourg: Fribourg University Press, 1959).

[12] Roger Bacon, *The Opus Majus*, ii. 611.

[13] A. C. Crombie, 'Grosseteste's Position in the History of Science', in *Robert Grosseteste*, ed. D. A. Callus (Oxford: Clarendon Press, 1955), 118.

[14] Ibid. 118–19.

[15] Euclid, *Elements*, Book IX, Proposition 20.

[16] John Buridan, *Questions on the Eight Books of the Physics of Aristotle*, Book VIII, Question 12, reprinted in M. Clagett, *The Science of Mechanics in the Middle Ages* (Madison, Wis.: University of Wisconsin Press, 1959), 533.

[17] A. C. Crombie, *Robert Grosseteste*, 119–24.

[18] William of Ockham, *Summulae in Phys.*, III. 5–7, in *Ockham Studies and Selections*, trans. and ed. S. C. Tornay (La Salle, Ill.: Open Court Publishing Co., 1938), 170–1.

[19] *Duns Scotus: Philosophical Writings*, 106–9.

[20] Nicolaus of Autrecourt, 'Second Letter to Bernard of Arezzo', in *Medieval Philosophy*, ed. H. Shapiro (New York: The Modern Library, 1964), 516–20.

[21] J. R. Weinberg, *Nicolaus of Autrecourt* (Princeton, NJ: Princeton University Press, 1948), 69.

[22] Ibid. 96–7.

6

The Debate over Saving
the Appearances

NICOLAUS COPERNICUS (1473–1543) received a sinecure as canon at Frauenburg through the efforts of his influential uncle, the Bishop of Ermland. As a consequence, Copernicus was able to spend several years studying at Italian universities, and to pursue his project of reforming mathematical planetary astronomy. In the *De revolutionibus* (1543), Copernicus revised Ptolemy's mathematical models by eliminating equant points and by taking the sun to be (roughly) the centre of planetary motions.

JOHANNES KEPLER (1571–1630) was born in the Swabian city of Weil. He was of delicate constitution, and passed an unhappy childhood. Kepler found relief in his studies and his Protestant faith. At the University of Tübingen, Michael Maestlin interested him in the Copernican astronomy. The sun-centred system appealed to Kepler on aesthetic and theological grounds, and he devoted his life to the discovery of the mathematical harmony according to which God must have created the universe.

    In 1594 he accepted a position as teacher of mathematics in a Lutheran school at Graz. Two years later he published the *Mysterium Cosmographicum*, in which he stated his "nest of regular solids" theory of planetary distances. This work, like all his writings, displayed a Pythagorean commitment informed by Christian fervour. In 1600, partly to escape pressure from Catholics in Graz, Kepler went to Prague as assistant to the great observational astronomer Tycho Brahe. He eventually gained access to Tycho's observations, and for the most part tempered his enthusiasm for mathematical correlations with respect for the accuracy of Tycho's data. Kepler published the first two laws of planetary motion in *Astronomia Nova* (1609), and the third law in *De Harmonice Mundi* (1619).

Osiander on Mathematical Models and Physical Truth

The question of proper method in astronomy was still debated in the sixteenth century. The Lutheran theologian Andreas Osiander affirmed the tradition of saving the appearances in his Preface to Copernicus's *De revolutionibus*. Osiander argued that Copernicus was working in the tradition of those astronomers who freely invent mathematical models in order to predict the positions of the planets. Osiander declared that it does not matter whether the planets really do revolve around the sun. What counts is that Copernicus has been able to save the appearances on this assumption. In a letter to Copernicus, Osiander tried to persuade him to present his sun-centred system as a mere hypothesis for which only mathematical truth was claimed.

Copernicus's Pythagorean Commitment

Copernicus, however, did not subscribe to this approach to astronomy. As a committed Pythagorean, he sought mathematical harmonies in phenomena because he believed they were "really there". Copernicus believed that his sun-centred system was more than a computational device.

Copernicus recognized that the observed planetary motions could be deduced with about the same degree of accuracy from his system, or from Ptolemy's system. Hence he acknowledged that selection of one of these competing models was based on considerations other than successful fit. Copernicus argued for the superiority of his own system by appealing to "conceptual integration" as a criterion of acceptability. He contrasted his own unified model of the solar system with Ptolemy's collection of separate models, one for each planet. He noted, moreover, that the sun-centred system explains the magnitudes and frequencies of the retrograde motions of the planets. The sun-centred system implies, for instance, that Jupiter's retrograde motion is more pronounced than that of Saturn, and that the frequency with which retrogression occurs is greater for Saturn than for Jupiter.* By contrast, Ptolemy's Earth-centred system provides no explanation of these facts.[1]

Copernicus died before having a chance to respond to Osiander's Preface to his book. Consequently, the sixteenth-century confrontation of the two methodological orientations—Pythagoreanism and the concern to save appearances—was not as sharp as it might have been.

Bellarmine v. Galileo

It remained for Cardinal Bellarmine and Galileo to state the rival positions with maximum intensity. Bellarmine informed Galileo in 1615 that it was permissible, from the standpoint of the Church, to discuss the Copernican system as a mathematical model to save the appearances. He indicated, moreover, that it is permissible to judge that the Copernican model is better able to save the appearances than is the Ptolemaic model. But Bellarmine insisted that to judge one mathematical model superior to another is not the same thing as to demonstrate the physical truth of the assumptions of the model.

The Jesuit mathematician Christopher Clavius had declared (in 1581) that Copernicus had saved the appearances of planetary motions by deducing theorems about them from *false* axioms. Clavius held that there was nothing exceptional about Copernicus's achievement, for, given a true theorem, any number of sets of false premises can be found which imply the theorem. Clavius himself preferred the Ptolemaic system, because he believed that an Earth-centred system is consistent with both the principles of physics and the teachings of the Church.

Bellarmine was aware that many influential churchmen shared

---

* Assuming, of course, that the orbital velocities of the planets decrease regularly, proceeding outwards from Mercury to Saturn.

the opinion of Clavius, and he warned Galileo that it would be dangerous to defend the position that the sun really is stationary, and that the Earth really does revolve around it.

Galileo, as is well known, overplayed his hand. Despite his disclaimers to the contrary, his *Dialogue Concerning the Two Great World Systems* was a thinly veiled polemic on behalf of the Copernican system. Galileo did not regard the heliocentric hypothesis as a mere computational device to save appearances. Indeed, he advanced a number of arguments in favour of the *physical truth* of the Copernican system. It was of great importance for the subsequent development of science that Galileo supplemented his Pythagorean commitment with the conviction that suitably chosen experiments can establish the existence of mathematical harmonies in the universe.

Kepler's Nest of Regular Solids

Kepler's Pythagorean Commitment

The Pythagorean orientation yielded substantial dividends in the astronomical investigations of Johannes Kepler. Kepler believed it to be significant that there exist just six planets and just five regular solids. Because he believed that God created the solar system according to a mathematical pattern, he sought to correlate the distances of the planets from the sun with these geometrical figures. In the *Mysterium Cosmographicum*, a book published in 1596, he announced with some pride that he had succeeded in gaining insight into God's plan of creation. Kepler showed that the distances of the planets can be correlated with the radii of spherical shells, which are inscribed within, and circumscribed around, a nest of the five regular solids. Kepler's arrangement was:

> Sphere of Saturn
> Cube
> Sphere of Jupiter
> Tetrahedron
> Sphere of Mars
> Dodecahedron
> Sphere of Earth
> Icosahedron
> Sphere of Venus
> Octahedron
> Sphere of Mercury

Kepler was able to achieve a rough agreement between the observed ratios of the radii of the planets and ratios calculated from the geometry of the nest of regular solids. However, he took values of planetary radii from data of Copernicus, which referred planetary distances to the centre of the Earth's orbit. Kepler hoped to improve the rough correlation achieved by his theory by referring planetary distances to the sun, thereby taking account of the eccentricity of the Earth's orbit. He recomputed the ratios of the planetary radii on this basis, using Tycho Brahe's more accurate data, and found that these ratios differed substantially from the ratios calculated from the regular-solid theory. Kepler accepted this as a refutation of his theory, but his Pythagorean faith was unshaken. He was convinced that the discrepancies between observation and theory themselves must be a manifestation of yet-to-be-discovered mathematical harmonies.

Kepler persevered in the search for mathematical regularities in the solar system, and eventually succeeded in formulating three laws of planetary motion:

(1) The orbit of a planet is an ellipse with the sun at one focus.
(2) The radius vector from the sun to a planet sweeps over equal areas in equal times.
(3) The ratio of the periods of any two planets is directly proportional to the ratio of the cubes of their mean distances from the sun.

Kepler's discovery of the Third Law is a striking application of Pythagorean principles. He was convinced that there must be a mathematical correlation between planetary distances and orbital velocities. He discovered the Third Law only after having tried a number of possible algebraic relations.

The committed Pythagorean believes that if a mathematical relation fits phenomena, this can hardly be a coincidence. But Kepler, in particular, formulated a number of mathematical correlations whose status is suspect. For example, he correlated planetary distances and their "densities". He suggested that the densities of the planets are inversely proportional to the square roots of their distances from the sun. Kepler had no way to determine independently the densities of the planets. In spite of this, he noted that the densities calculated from this mathematical relation could be correlated with the densities of well-known terrestrial substances. He set up the following table:[2]

Kepler's Distance—Density Relation

Planet	$Density = 1\sqrt{distance}$ (Earth = 1,000)	Terrestrial substance
Saturn	324	The hardest precious stones
Jupiter	438	The lodestone
Mars	810	Iron
Earth	1,000	Silver
Venus	1,175	Lead
Mercury	1,605	Quicksilver

Kepler noted with satisfaction that it would be appropriate to correlate the sun with gold, the density of which is greater than that of quicksilver. Of course, Kepler did not believe that the Earth was composed of silver and Venus of lead, but he did believe it important that his calculated planetary densities correspond to the densities of these terrestrial substances.

From the Pythagorean standpoint, the adequacy of a mathematical correlation is determined by appeal to the criteria of "successful fit" and "simplicity". Provided that a relation is not unduly complex mathematically, if it fits the phenomena under consideration, it must be important. But a person who does not share the Pythagorean faith doubtless would judge Kepler's distance–density correlation to be a coincidence. Such a person might appeal to criteria other than successful fit and simplicity, on the grounds that application of these criteria alone is not sufficient to distinguish genuine correlations from coincidental correlations.

Bode's Law

The evaluation of mathematical correlations has been a continuing problem in the history of science. In 1772, for example, Johann Titius suggested a correlation that was in the Pythagorean tradition. He noted that the distances of the planets from the sun could be correlated with the "suitably adjusted" terms of the geometrical series 3, 6, 12, 24 . . . , viz.:

Bode's Law

	4	4	4	4	4
	0	3	6	12	24
Calculated	4	7	10	16	28
Planet	Mercury	Venus	Earth	Mars	(Asteroids)
Observed	3.9	7.2	10	15.2	

	4	4	4	4	
	48	96	192	384	
Calculated	52	100	196	388	
Planet	Jupiter	Saturn	(Uranus)	(Neptune)	(Pluto)
Observed	52.0	95.4	191.9	300.7	395

The numbers thus obtained are in striking agreement with the observed distances, relative to Earth = 10. The noted astronomer

Johann Bode was greatly impressed by this relation. He accepted the Pythagorean position that a successful fit is not likely to be a coincidence. Because he championed this relation, it came to be known as 'Bode's Law'. In 1780, an astronomer's judgement of the significance of Bode's Law was a good measure of the strength of his commitment to the Pythagorean orientation.

Then, in 1781, William Herschel discovered a planet beyond Saturn. Astronomers on the continent calculated the distance of Uranus from the sun and found it to be in excellent agreement with the next term in Bode's Law (196). Eyebrows were raised. The sceptics no longer could dismiss this correlation as an "after the fact" numerical coincidence. An increasing number of astronomers began to take Bode's Law seriously. A search was undertaken for the "missing planet" between Mars and Jupiter, and the asteroids Ceres and Pallas were discovered in 1801 and 1802. Although the asteroids were much smaller than Mercury, their distances were such that astronomers who believed in Bode's Law were satisfied that the missing term in the series had been filled.

After it became apparent that the motion of Uranus was being affected by a still more distant planet, J. C. Adams and U. J. J. Leverrier independently calculated the position of this new planet. One ingredient in their calculations was the assumption that the mean distance of the new planet would be given by the next term in Bode's Law (388). The planet Neptune was discovered by Galle in the region predicted by Leverrier. However, continued observation of the planet revealed that its mean distance from the sun (relative to Earth = 10) is about 300, which is not in good agreement with Bode's Law.*

With the inclusion of Neptune, Bode's Law no longer satisfied the criterion of successful fit. Hence one may be a Pythagorean today without being impressed by Bode's Law. On the other hand, since Pluto's distance is very close to the Bode's Law value for the next planet beyond Uranus, a person with a Pythagorean bent might be tempted to explain away the anomalous case of Neptune by insisting that Neptune is a lately captured acquisition of the solar system, and not one of the original planets at all.

---

* Neptune's position in its orbit at the time of discovery was such that the over-estimation of its distance from the sun did not greatly affect the accuracy of the prediction of its position against the background stars.

Notes

[1] Copernicus, *On the Revolutions of the Heavenly Spheres*, bk. 1, chap. 10.
[2] Kepler, *Epitome of Copernican Astronomy*, trans. C. G. Wallis, in *Ptolemy, Copernicus, Kepler*—Great Books of the Western World, vol. 16 (Chicago: Encyclopaedia Britannica, Inc. 1952), 882.

7

The Seventeenth-Century
Attack on Aristotelian Philosophy

I. GALILEO

GALILEO GALILEI (1564–1642) was born at Pisa, of noble but impov-
erished parents. In 1581 he enrolled at the University of Pisa to pursue
the study of medicine, but soon abandoned his medical studies in favour
of mathematics and physics.

In 1592 he was appointed Professor of Mathematics at the University
of Padua, where he remained until 1610. During this period, Galileo
made important telescopic observations of sunspots, the surface of the
moon, and four of the satellites of Jupiter. These observations were
inconsistent with implications of the Church-sanctioned Aristotelian
world-view, in which the celestial realm is immutable and the Earth is
the centre of all motion.

Galileo became mathematician-in-residence to the Grand Duke of
Tuscany in 1610. He engaged in a series of disputes with Jesuit and
Dominican philosophers, at one point lecturing these worthies on the
proper way to interpret the Scriptures so as to effect agreement with the
Copernican astronomy (*Letter to the Grand Duchess Christina*, 1615).

Galileo's admirer Maffeo Barberini was elected pope in 1623, and
Galileo sought and received permission to prepare an impartial study of

the rival Copernican and Ptolemaic systems. *The Dialogue Concerning the Two Chief World Systems* (1632) contained a preface and conclusion which indicated that the rival systems are mere mathematical hypotheses to save the appearances. The remainder of the book, which Galileo wrote in Italian to reach a wider audience, contained numerous arguments for the physical truth of the Copernican alternative.

Galileo was called before the Inquisition and forced to abjure his errors. He retired to Florence under the watchful eyes of his enemies. However, he gained revenge with the publication of the *Dialogues Concerning Two New Sciences* (1638), which demonstrated the inadequacy of Aristotle's physics, thereby removing a major support of geocentrism.

The Pythagorean Orientation and the Demarcation of Physics

Galileo was convinced that the book of nature is written in the language of mathematics. For this reason, he sought to restrict the scope of physics to assertions about "primary qualities". Primary qualities are qualities essential to the very concept of body. Galileo believed that primary qualities such as shape, size, number, position, and "quantity of motion", are objective properties of bodies, and that secondary qualities, such as colours, tastes, odours, and sounds, exist only in the mind of the perceiving subject.[1]

By restricting the subject-matter of physics to primary qualities and their relations, Galileo excluded teleological explanations from the range of permissible discourse of physics. According to Galileo, it is not a bona fide scientific explanation to state that a motion takes place in *order that* some future state may be realized. In particular, he urged that Aristotelian interpretations in terms of "natural motions" towards "natural places" do not qualify as scientific explanations. Galileo realized that he could not prove false an assertion such as "unsupported bodies move toward the Earth in order to reach their 'natural place'." But he also realized that this type of interpretation can be excluded from physics because it fails to "explain" the phenomena.

Implicit in Galileo's analysis is a distinction between two stages in the evaluation of interpretations in science. The first stage is to demarcate scientific interpretations from non-scientific interpretations. Galileo agreed with Aristotle that this is a question of

circumscribing the proper subject-matter of science. The second
stage is to determine the acceptability of those interpretations
that do qualify as scientific. Galileo's approach to the problem of
evaluating interpretations in science may be represented as follows:

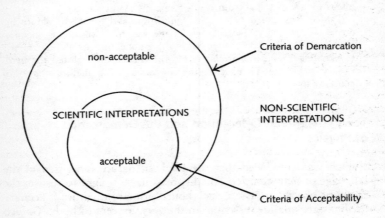

Galileo established the circumference of the larger circle by restricting the subject-matter of physics to statements about primary
qualities.

One consequence of Galileo's demarcation of physics is that
the motions of bodies are described with respect to a system of
coordinates in space. Galileo replaced Aristotle's qualitatively differentiated space by a quantitatively differentiated geometrical
space.

But his break with the qualitatively differentiated space of the
Aristotelian universe was never complete. In the early work *De
Motu*, Galileo himself affirmed the doctrine of "natural places".[2]
Although subsequently he sought to exclude interpretations in
terms of "natural places" from physics, he remained committed
throughout his life to the doctrine that only circular motion is
suited to celestial bodies. Galileo believed that the Earth itself is a
bona fide celestial body, and he attempted to prove to the Aristotelians that the Earth, and bodies on its surface, participate in
the perfection of circular motion. For example, he maintained that,
in the absence of all resistance, motion along the Earth's surface

would persist undiminished indefinitely.[3] In this instance, Galileo was guilty of formulating the same type of interpretation that his demarcation of physics was intended to exclude.

Theory of Scientific Procedure

Galileo's anti-Aristotelian polemic was *not* directed against Aristotle's inductive—deductive method. He accepted Aristotle's view of scientific inquiry as a two-stage progression from observations to general principles and back to observations.

Moreover, Galileo approved Aristotle's position that explanatory principles must be induced from the data of sense experience. In this regard, Galileo observed that Aristotle himself would have repudiated the doctrine of the immutability of the heavens had he been in possession of the seventeenth-century telescopic evidence on sunspots. He declared that "it is better Aristotelian philosophy to say 'Heaven is alterable because my senses tell me so', than to say 'Heaven is inalterable because Aristotle was so persuaded by reasoning.'"[4]

Galileo's remarks about scientific procedure were directed against practitioners of a false Aristotelianism, who short-circuited the Method of Resolution and Composition by beginning, not with induction from sense experience, but with Aristotle's own first principles. This false Aristotelianism encouraged a dogmatic theorizing which cut off science from its empirical base. Galileo frequently condemned this perversion of Aristotle's methodology.

The Method of Resolution

Galileo insisted on the importance to physics of abstraction and idealization, thereby extending the reach of inductive techniques. In his own work, he made use of idealizations such as 'free fall in a vacuum' and the 'ideal pendulum'. These idealizations are not exemplified directly in phenomena. They are formulated by extrapolating from serially ordered phenomena. The concept free fall in a vacuum, for example, is an extrapolation from the observed behaviour of bodies dropped in a series of fluids of decreasing density.[5] The concept ideal pendulum is likewise an idealization. An "ideal" pendulum is one whose bob is attached to a "mass-less" string in which there are no frictional forces due to different periods of motion for different segments of the string. Moreover, the motion of such a pendulum is unimpeded by air resistance.

Galileo's work in mechanics testifies to the fertility of these concepts. He was able to deduce the approximate behaviour of falling bodies and real pendulums from explanatory principles that specify properties of idealized motions. One important consequence of this use of idealizations was to emphasize the role of creative imagination in the Method of Resolution. Hypotheses about idealizations can be obtained neither by induction by simple enumeration nor by the methods of agreement and difference. It is necessary for the scientist to intuit which properties of phenomena are the proper basis for idealization, and which properties may be ignored.[6]

The Method of Composition

Grosseteste and Roger Bacon had augmented the Method of Composition by suggesting the deduction of consequences not included in the data initially used to induce explanatory principles. Galileo made a striking application of this procedure by deducing from his hypothesis of the parabolic trajectory of projectiles, that the maximum range is achieved at 45 degrees. That the maximum range is achieved at 45 degrees was known prior to Galileo's work. Galileo's achievement was an explanation of this fact. Galileo also deduced from the parabolic trajectory that the same range is achieved for angles of elevation equally far removed from 45 degrees, e.g. 40 degrees and 50 degrees. He claimed that this had not been recognized by gunners, and used this occasion to eulogize the superiority of mathematical demonstration over untutored experience.[7]

Experimental Confirmation

Grosseteste and Roger Bacon had appended to the Method of Resolution and Composition a third stage in which the conclusions reached are further tested experimentally. Galileo's attitude toward this third stage has received very different evaluations. He has been hailed as a champion of experimental methodology. But he also has been criticized for failing to appreciate the importance of experimental confirmation. A case can be made for each evaluation, both from his comments on scientific procedure and from his scientific practice.

Galileo made ambivalent pronouncements on the value of experimental confirmation. His dominant emphasis is affirmative. For example, in the *Dialogues Concerning Two New Sciences*, after Salviati had deduced the law of falling bodies, Simplicio demanded experimental confirmation of this relation. Galileo had Salviati reply that "the request that you, as a man of science, make, is a very reasonable one; for this is the custom—and properly so—in those sciences where mathematical demonstrations are applied to natural phenomena."[8]

However, it is also true that Galileo occasionally wrote as if experimental confirmation were relatively unimportant. For example, after having deduced the variation of range of a projectile with the angle of elevation, he wrote that "the knowledge of a single fact acquired through a discovery of its causes prepares the mind to understand and ascertain other facts without need of recourse to experiment."[9]

A similar ambivalence over experimentation is found in Galileo's scientific practice. Very often he described experiments which he probably had performed himself.

From the standpoint of the history of physics, Galileo's most important experiments were on the problem of falling bodies. Galileo reported that he had confirmed the law of falling bodies by rolling balls down inclined planes of various heights. Although he did not state the values obtained in these experiments, he did go into considerable detail about the construction of the planes and the measurement of the time of fall by a water clock.[10]

Galileo also reported that he had performed experiments with a pendulum to confirm the hypothesis that the speeds achieved by a body moving down planes of different inclinations are equal when the heights of the planes are equal. He claimed that if the motion of a pendulum, consisting of a bullet tied to a string, is arrested when the string strikes a nail, then the bullet reaches the same height as it did when its oscillation was unimpeded.

Galileo maintained that the pendulum–nail experiment indirectly confirmed the hypothesis about motion on inclined planes. He noted that a direct confirmation by rolling a ball down one plane and up a second is impractical because of the "obstacle" at the point of junction.[11]

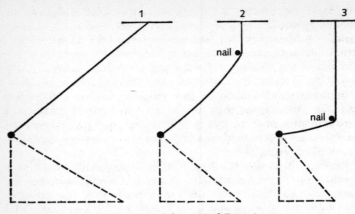

Galileo's Pendulum–Nail Experiment

Galileo's less widely publicized experiments include a demonstration that a floating hollow wooden vessel does not sink when the cavity is filled with water,[12] and an occultation of stars by a rope to show that stellar diameters are exaggerated by the naked eye.[13]

In spite of his descriptions of experiments supposedly performed, however, Galileo's commitment to experimental confirmation was not complete. There are instances in which he dismissed experimental evidence that seemed to count against his theories.

In the early work *De motu*, for example, Galileo formulated the relationship $\frac{v_1}{v_2} = \frac{d_1 - d_m}{d_2 - d_m}$, in which v_1 and v_2 are the velocities of fall of two spheres of equal volume through a medium, d_1 and d_2 are the densities of these bodies, and d_m is the density of the medium. Commenting on this relationship, he admitted that if one drops from a tower two balls, chosen so that $\frac{d_1 - d_m}{d_2 - d_m} = 2$, a corresponding ratio in velocities is *not* observed. Indeed, the two balls hit the ground at approximately the same time. Galileo attributed this failure in confirmation to "unnatural accidents".[14] In this instance, he was anxious to recommend a mathematical relationship, which he believed followed from Archimedes' law of buoyancy, in spite of the fact that it does not describe the behaviour of bodies falling through air. Galileo subsequently abandoned this relationship in favour of a kinematic approach in which the distance of fall is related to the elapsed time.

Galileo also dismissed evidence that was unfavourable to his theory of the tides. He believed that the tides are caused by the periodic reinforcement and opposition of two motions of the Earth—its annual revolution around the sun and its daily rotation on its axis. Galileo's hypothesis, roughly put, was that for a given port *P*, revolution and rotation augment one another at midnight and oppose one another at noon.

The result of this periodic reinforcement and cancellation is that the water offshore is left behind at night and is piled up along the

Galileo's Theory of the Tides

coast in the day-time. It follows from Galileo's theory that there should be just one high tide each day at a given location, and that it should occur around noon.

But it was a well-established fact that there are two high tides per day at a given port. Moreover, the times at which they occur vary around the clock from day to day.

Galileo attributed the divergence of theory and fact to the operation of "secondary causes", such as the irregular depth of the sea and the shape and orientation of the coastline. Galileo maintained that the very fact that there are tides at all provides support for the Copernican theory. He was so anxious to find arguments for the twofold motion of the Earth that he was willing to explain away evidence that counted against his theory of tidal action.

In addition, there is one instance in which Galileo reported having confirmed a law for a range in which the law does not hold. He claimed to have observed that the period of a pendulum is independent of the amplitude of its swings for angles as high as 80 degrees from the perpendicular.[15] But the period of a pendulum is independent of its amplitude only for small displacements from the perpendicular. One must conclude either that Galileo did not bother to experiment with swings of large angle, or that his observations were extremely careless. Perhaps his error may be attributed to a strong conviction about how a pendulum *should* swing.

The Ideal of Deductive Systematization

Galileo affirmed the Archimedean ideal of deductive systematization. And he also accepted the Platonic distinction between the real and the phenomenal, with which this ideal often was associated. From the standpoint of this distinction, it is natural to de-emphasize discrepancies between the theorems of deductive systems and what actually is observed. Such discrepancies may be attributed to "unimportant" experimental complications. As noted above, Galileo sometimes took recourse to this approach.

However, a more important aspect of Galileo's Archimedean–Platonic commitment was his emphasis on the value of abstraction and idealization in science. This was the converse side, as it were, of his willingness to explain away discrepancies between theory and observation. It was stressed above that much of Galileo's success in physics may be attributed to his ability to bracket out various empirical complications in order to work with ideal concepts such as "free fall in a vacuum", "ideal pendulum", and the "frictionless motion of a ship through the ocean". This is a positive feature of the ideal of deductive systematization. Galileo himself was quite sophisticated about the role of abstraction in science. He wrote that

just as the computer who wants his calculations to deal with sugar, silk, and wool must discount the boxes, bales, and other packings, so the mathematical scientist, when he wants to recognize in the concrete the effects which he has proved in the abstract, must deduct the material hindrances, and if he is able to do so, I assure you that things are in no less agreement than arithmetical computations. The errors, then, lie not in the abstractness or concreteness, not in geometry or physics, but in a calculator who does not know how to make a true accounting.[16]

Notes

[1] Galileo, *The Assayer*, trans. S. Drake, in *The Controversy on the Comets of 1618*, trans. S. Drake and C. D. O'Malley (Philadelphia: University of Pennsylvania Press, 1960), 309.

[2] Galileo, *On Motion*, trans. I. E. Drabkin, in Galileo, *On Motion and On Mechanics*, trans. I. E. Drabkin and S. Drake (Madison, Wis.: The University of Wisconsin Press, 1960), 14–16.

[3] Galileo, *Dialogue Concerning the Two Chief World Systems*, trans. S. Drake (Berkeley, Calif.: University of California Press, 1953), 148;

Dialogues Concerning Two New Sciences, trans. H. Crew and A. de Salvio (New York: Dover Publications, 1914), 181–2;

"Second Letter from Galileo to Mark Welser on Sunspots", in *Discoveries and Opinions of Galileo*, trans. and ed. S. Drake (Garden City, NY: Doubleday Anchor Books, 1957), 113–14.

[4] Galileo, *Two World Systems*, 56.

[5] Galileo, *Two New Sciences*, 72.

[6] Galileo, *Two World Systems*, 207–8.

[7] Galileo, *Two New Sciences*, 276.

[8] Ibid. 178.

[9] Ibid. 276.

[10] Ibid. 178–9.

[11] Ibid. 172.

[12] Galileo, *Discourse on Bodies in Water*, trans. T. Salusbury (Urbana, Ill.: University of Illinois Press, 1960), 22.

[13] Galileo, *Two World Systems*, 361–4.

[14] Galileo, *On Motion*, 37–8.

[15] Galileo, *Two New Sciences*, 254–5, 85; *Two World Systems*, 450.

[16] Galileo *Two World Systems*, 207–8.

II. FRANCIS BACON

FRANCIS BACON (1561–1626) was a son of Sir Nicholas Bacon, Lord
Keeper to Queen Elizabeth I. Bacon entered Trinity College, Cambridge
at the age of thirteen, and there developed an antipathy towards Aristotelian philosophy. Subsequently, he studied law at Gray's Inn and was
admitted to the bar in 1586.

Bacon made numerous efforts to secure a governmental appointment
from the Queen, but although his uncle William Cecil, later Lord Burgh-
ley, was Elizabeth's most important minister, the appointment was not
forthcoming. This doubtless was due in part to Bacon's defence of the
rights of Commons against certain proposals urged by the Queen's
ministers.

Following the accession of James I, Bacon's fortunes soared. He was
knighted in 1603, became Attorney-General in 1613, Lord Keeper in
1617, Lord Chancellor in 1618, Baron Verulam in 1618, and Viscount
St Albans in 1621. Shortly thereafter, he pleaded guilty to taking gifts
from persons with cases before him in his capacity as Lord Chancellor.
Bacon insisted that he had not allowed the receipt of gifts to influence
his judgements in these cases, but he offered no defence against the
charge that he had accepted the gifts. Bacon was fined, jailed, and
banished from public life by his peers in the House of Lords, but the
King remitted the fine and terminated his imprisonment after a few days.

Bacon spent much of his time during the last five years of his life
working on his *Great Instauration*, a proposed reformulation of the
sciences. His most important contribution towards this Instauration was
the *Novum Organum*, which he had published in 1620. In this work, he
outlined a "new" scientific method to replace that of Aristotle. He also
created an influential image of co-operative scientific inquiry in the *New
Atlantis* (1627).

The Controversy over the Value of Bacon's Contribution

Francis Bacon is a controversial figure in the history of science.
In the eyes of the founders of the Royal Society he was the
prophet of a new scientific methodology. The *philosophes* likewise
regarded Bacon to be an innovator, a champion of a new inductive–
experimental method. But Alexandre Koyré and E. J. Dijksterhuis,
two eminent twentieth-century historians, have minimized the
value of Bacon's contributions. They have emphasized that Bacon

achieved no new results in science, and that his criticism of Aristotelian method was neither original nor incisive. According to Dijksterhuis, Bacon's role in science was analogous to the military role of the lame Greek poet Tyrtaeus. Tyrtaeus could not fight, but his war-songs brought inspiration to those who could.[1]

The disputants agree about several aspects of Bacon's contribution: (1) that Bacon himself did not enrich science by means of concrete examples of his professed method; (2) that Bacon's great literary gifts enabled him to express his ideas so effectively that many scholars have attributed to him a large role in the scientific revolution of the seventeenth century; and (3) that Bacon's originality, if any, is his theory of scientific method.

Bacon himself claimed originality for his method. He chose as title of his principal work on method '*Novum Organum*', thereby indicating that his method was to replace the method discussed in the *Organon*, a medieval compilation of Aristotle's writings. Some critics have maintained that Bacon was successful. For instance, John Herschel declared in his influential *Preliminary Discourse on Natural Philosophy* (1830) that

by the discoveries of Copernicus, Kepler, and Galileo, the errors of the Aristotelian philosophy were effectually overturned on a plain appeal to the facts of nature; but it remained to show on broad and general principles, how and why Aristotle was in the wrong; to set in evidence the peculiar weakness of his method of philosophizing, and to substitute in its place a stronger and better. This important task was executed by Francis Bacon.[2]

Criticism of Aristotelian Method

But was Bacon's method a "new" *Organon*? Bacon insisted that the first requirement of scientific method is that the natural philosopher should purge himself of prejudices and predispositions in order to become again as a child before nature. He noted that the study of nature has been obscured by four classes of "Idols" which beset men's minds. Idols of the Tribe have their foundation in human nature itself. The understanding is prone to postulate more regularity in nature than it actually finds, to generalize hastily, and to overemphasize the value of confirming instances. Idols of the Cave, by contrast, are attitudes towards experience that arise from the upbringing and education of men as individuals. Idols of the Market-Place are distortions that ensue when the meanings of words are reduced to the lowest common denominator of vulgar usage, thereby impeding scientific concept-formation. And Idols of the Theatre are the received dogmas and methods of the various philosophies.

Aristotle's philosophy was an Idol of the Theatre that Bacon was most anxious to discredit. It must be emphasized, however, that Bacon accepted the main outline of Aristotle's inductive–deductive theory of scientific procedure. Bacon, like Aristotle, viewed science as a progression from observations to general principles and back to observations. It is true that Bacon emphasized the inductive stage of scientific procedure. But he did assign to deductive arguments an important role in the confirmation of inductive generalizations.[3] Moreover, Bacon insisted that the fruits of scientific inquiry are new works and inventions, and noted that this is a matter of deducing from general principles consequences that have practical application.[4]

But although Bacon did accept Aristotle's theory of scientific procedure, he was highly critical of the way in which this procedure had been carried out. With respect to the inductive stage, Bacon issued a three-part indictment.

First, Aristotle and his followers practise a haphazard, uncritical collection of data. Francis Bacon called for a thoroughgoing implementation of Roger Bacon's Second Prerogative of Experimental Science, viz., the use of systematic experimentation to gain new knowledge of nature. In this connection, Francis Bacon stressed the value of scientific instruments in the collection of data.

Second, the Aristotelians generalize too hastily. Given a few observations, they leap at once to the most general principles, and then use these principles to deduce generalizations of lesser scope.

Third, Aristotle and his followers rely on induction by simple enumeration, in which correlations of properties found to hold for several individuals of a given type, are affirmed to hold for all individuals of that type. But application of this inductive technique often leads to false conclusions, because negative instances are not taken into account (Bacon did not mention the emphasis placed on a method of difference by such medieval writers as Grosseteste and Ockham).

With respect to the deductive stage of scientific inquiry, Bacon made two principal complaints. Bacon's first complaint was that the Aristotelians had failed to define adequately such important predicates as 'attraction', 'generation', 'element', 'heavy', and 'moist', thereby rendering useless those syllogistic arguments in which these predicates occur.[5] Bacon correctly pointed out that syllogistic demonstration from first principles is effective only if the terms of the syllogisms are well defined.

Bacon's second complaint was that Aristotle and his followers had reduced science to deductive logic by overemphasizing the deduction of consequences from first principles. Bacon stressed that deductive arguments are of scientific value only if their premisses have proper inductive support.

At this point, Bacon should have distinguished Aristotle's theory of procedure from the way in which this theory of procedure had been misappropriated by some subsequent thinkers who called themselves "Aristotelians". Practitioners of a false Aristotelianism had short-circuited Aristotle's method by beginning, not with induction from observational evidence, but with Aristotle's own first principles. This false Aristotelianism encouraged a dogmatic theorizing by cutting off science from its empirical base. But Aristotle himself had insisted that first principles be induced from observational evidence. Bacon was unfair to condemn Aristotle for reducing science to deductive logic.

"Correction" of Aristotelian Method

Bacon put forward his "new" method for science in order to overcome the supposed deficiencies of the Aristotelian theory of procedure. The two principal features of Bacon's new method were an emphasis on gradual, progressive inductions, and a method of exclusion.

Bacon believed that properly conducted scientific inquiry is a step-by-step ascent from the base to the apex of a pyramid of propositions.

Bacon suggested that a series of "natural and experimental histories" should be compiled in order to establish a secure base for the pyramid. Bacon himself contributed works on the winds, the ebb and flow of the tides, and the longevity and modes of life of various peoples and animals. Unfortunately, he took much

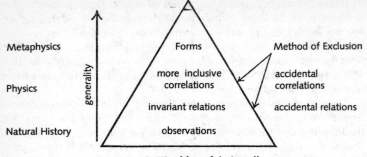

Bacon's *"Ladder of Axioms"*

of the materials for his natural histories from untrustworthy sources.

Bacon held that, after having established the facts in a particular science, the natural philosopher should seek correlations within these facts. And he insisted on a gradual inductive ascent, from correlations of a low degree of generality to those which are more inclusive.

Bacon was aware that some correlations among facts are only "accidental" correlations. To weed out accidental correlations, he formulated a method of exclusion. Bacon suggested that accidental correlations often may be identified by inspecting Tables of Presence, Absence, and Degrees. Any correlation for which there is an instance in which one attribute is absent when another is present, or instances in which one attribute decreases when the other increases, is to be excluded from the pyramid. Bacon believed that after accidental correlations had been excluded in this way, only essential correlations would remain. And essential correlations are suitable subject-matter for further inductive generalization.

Bacon cited the method of exclusion as an important point of superiority of his method over that of Aristotle. He correctly maintained that simple enumeration, which was one of the inductive procedures employed by Aristotle, is inadequate to distinguish essential correlations from accidental correlations. Bacon claimed that application of the method of exclusion can effect this distinction, because this method places due weight on absence and relative intensity.

Bacon was sufficiently realistic to recognize that, in many cases, it is difficult to find essential correlations merely by inspecting Tables of Presence, Absence, and Degrees. For this reason, he singled out various types of "Prerogative Instances" which are of special value in the search for essential correlations. He seemed to have believed that it is of the very nature of these instances to reveal essential correlations.

Perhaps the most important of Bacon's 27 Prerogative Instances is the "Instance of the Fingerpost". An Instance of the Fingerpost is an instance that decides the issue between competing explanations. Bacon himself suggested a crucial instance of this type to decide between two hypotheses about the ebb and flow of the tides. The first hypothesis was that the tides are an advance and retreat of waters, on analogy to water rocked to-and-fro in a basin. The second hypothesis was that the tides are a periodic lifting and falling of waters. Bacon noted that the basin hypothesis would be falsified if it could be shown that the temporally coincident high tides on the shores of Spain and Florida were not accompanied by ebb tides elsewhere. He suggested that a study of tides on the coasts of Peru and China would settle the issue.[6]

Bacon recognized that an instance is "crucial" only if it is inconsistent with every set of explanatory premises save one. But it is not possible to prove that a statement about a type of phenomena can be deduced from just these several sets of premises, and no others. Bacon was guilty of overestimating the logical force of Instances of the Fingerpost. Nevertheless, the elimination of hypotheses whose deductive consequences (given specific antecedent conditions), are not in agreement with observations, may be of value in the search for a more adequate explanation. Of course, Francis Bacon did not invent this method of falsification. Aristotle had employed it, and Grosseteste and Roger Bacon had recommended this method as a standard way to establish a hypothesis by eliminating competing hypotheses.

The Search for Forms

Bacon referred to the most general principles at the apex of the pyramid as "Forms". Forms are the verbal expressions of relations among "simple natures", those irreducible qualities present in the objects we perceive. Bacon believed that various combinations of these simple natures constitute the objects of our experience, and that if we could but gain knowledge of Forms, it would be possible to control and modify the forces of nature.

In certain of his comments about Forms, Bacon seems to have conceived the union of simple natures in terms of an alchemical analogy. For instance, he declared that

> he who knows the forms of yellow, weight, ductility, fixity, fluidity, solution, and so on, and the methods for superinducing them, and their gradations and modes, will make it his care to have them joined together in some body, whence may follow the transformation of that body into gold.[7]

Bacon himself contributed inquiries into the Forms of heat, whiteness, the attraction of bodies, weight, taste, memory, and the "Spirit enclosed within tangible bodies".[8]

Bacon's Forms are neither Platonic forms nor Aristotelian formal causes. Rather, Forms supposedly express those relations among physical properties that have the power to produce effects. In Aristotelian terms, Bacon's Forms refer to the material and efficient aspects of causation, as well as to the merely formal aspect.

In many cases (magnetism and the "Spirit enclosed within tangible bodies" are exceptions), Bacon specified Forms in terms of the configurations and motions of the invisible parts of bodies. He accepted the atomist principle that macroscopic effects are to be explained by submacroscopic interactions. But he did not accept the atomists' position that impact and impenetrability are the fundamental properties of atoms. Bacon attributed to the parts of bodies "forces" and "sympathies". Moreover, he did not accept the idea of a continuous void through which the atoms are dispersed.

Bacon placed two requirements on Forms: these propositions must be true in every instance, and the converses of these propositions must be true as well.* Bacon's Form of heat, for instance, states an identity of 'heat' and 'a rapid expansive motion of the small particles of bodies, which particles are restrained from escaping from the body's surface'.[10] According to Bacon, if heat is

---

* These requirements correspond to Peter Ramus's Rules of Truth and Wisdom, respectively.[9]

present, then so is this rapid expansive motion, and conversely. A similar convertibility supposedly holds for all Forms.

Bacon sometimes spoke of Forms as "laws". For example, in Book 2 of *Novum Organum*, he wrote that

when I speak of Forms, I mean nothing more than those laws and determinations of absolute actuality, which govern and constitute any simple nature, as heat, light, weight, in every kind of matter and subject that is susceptible of them. Thus the Form of Heat or the Form of Light is the same thing as the Law of Heat or the Law of Light.[11]

If extracted from context, certain of Bacon's remarks about "laws" have a modern ring. But several of Bacon's emphases are non-modern. In the first place, Bacon construed physical laws on the model of decrees enforced by a civil power. In the second place, Bacon was not interested in expressing laws in mathematical form. And in the third place, Bacon viewed the universe as a collection of substances which have properties and powers, and which stand in relations one to another. He did not view the universe as a flux of events which occur in lawful patterns. In this regard, Bacon's metaphysics is still Aristotelian.

One must conclude that Bacon's search for Forms is still very much in the Aristotelian tradition. John Herschel greatly overstated the case for the originality of Bacon's theory of procedure.

Bacon as Propagandist for Organized Scientific Research

But if this were all there was to say about Bacon, it would be difficult to understand why he is a controversial figure in the history of science. It is true that Bacon sought to reform scientific method. However, there is more to Bacon's vision of science than his suggested "corrections" of Aristotle's theory of procedure.

Bacon accepted as a moral imperative that man is to recover the dominion over nature which he lost in the Fall. He repeatedly emphasized that men must control and redirect natural forces so as to improve the quality of life of their fellow human beings. Thus the discovery of Forms is only the proximate goal of scientific inquiry. One must gain knowledge of Forms before one can coerce nature to serve human purposes. But the ultimate goal of scientific inquiry is power over nature. Bacon's emphasis on the practical application of scientific knowledge stands in marked contrast to Aristotle's position that knowledge of nature is an end in itself. It is this emphasis on the control of natural forces that most clearly sets apart Bacon's philosophy from the Aristotelian philosophy he hoped to overthrow.

This emphasis on the practical application of scientific knowledge accounts for much of Bacon's excessively hostile polemics against Aristotle. Farrington is correct to point out that Bacon's hostility reflects *moral* outrage—Aristotle's philosophy not only has not led to new works to benefit mankind, but also has thwarted those few attempts that have been made.[12] By contrast, Bacon extolled the progress that had been made in the various craft traditions, and cited the inventions of printing, gunpowder, and the mariner's compass as examples of what can be accomplished by men not under the spell of Idols of the Theatre.

An important aspect of Bacon's new vision of science is that the recovery of man's dominion over nature is possible only through co-operative inquiry. In the service of this conviction, Bacon launched numerous attempts to introduce reforms administratively. He directed his appeals for support of co-operative projects almost exclusively to the Crown and its ministers, rather than to the universities, a strategy which reflected his very low estimate of contemporary academic life. But he was not successful. His vision of co-operative inquiry reached fruition only in the succeeding generation, when the Royal Society undertook to implement, not only Bacon's general attitude toward science, but also a number of Bacon's specific projects.

A further aspect of Bacon's new view of science is the divorce effected between science on the one hand, and teleology and natural theology on the other hand. Bacon restricted inquiry into final causes to the volitional aspects of human behaviour, observing that the search for final causes of physical and biological phenomena leads to purely verbal disputes which impede scientific progress.[13] Bacon's exclusion of final causes from natural science reflects his insistence that the scientist become again a child before nature. To view nature through the prism of purposive adaptation, whether divinely ordained or not, is to fail to come to grips with nature on its own terms. Preoccupation with the question "for what purpose?" makes unlikely the discovery of Forms and the subsequent improvement of the human condition.

Notes

[1] E. J. Dijksterhuis, *The Mechanization of the World Picture*, trans. C. Dikshoorn (Oxford: Clarendon Press, 1961), 402.

[2] John F. W. Herschel, *A Preliminary Discourse on the Study of Natural Philosophy* (London: Longman, Rees, Orme, Brown and Green, and John Taylor, 1831), 113–14.

[3] Francis Bacon, *Novum Organum*, I, Aphorism CVI.

[4] Ibid., II, Aphorism X.

[5] F. Bacon, 'Plan of the Work', in *The Works of Francis Bacon*, viii, ed. J. Spedding, R. L. Ellis, and D. D. Heath (New York: Hurd and Houghton, 1870), 41; *Novum Organum*, I, Aphorism XV.

[6] F. Bacon, *Novum Organum*, II, Aphorism XXXVI.

[7] Ibid., II, Aphorism V.

[8] Ibid., II, Aphorisms XI–XXXVI.

[9] See Paolo Rossi, *Francis Bacon, From Magic to Science*, trans. S. Rabinovitch (London: Routledge & Kegan Paul, 1968), 195–8.

[10] F. Bacon, *Novum Organum*, II, Aphorism XX.

[11] Ibid., II, Aphorism XVII.

[12] See Benjamin Farrington, *The Philosophy of Francis Bacon* (Liverpool: Liverpool University Press, 1964), 30.

[13] F. Bacon, *Novum Organum*, II, Aphorism II.

III. DESCARTES

RENÉ DESCARTES (1596–1650) attended the Jesuit College at La Flèche and received a law degree from the University of Poitiers in 1616. But because he shared in a considerable family fortune, it was not necessary for him to practise law. Descartes was very much interested in mathematics, science, and philosophy, and he decided to combine intellectual pursuits with travel. He spent several years travelling about Europe, frequently in the capacity of gentleman volunteer in various armies.

In 1618 Descartes made the acquaintance of the physicist Isaac Beeckman, who encouraged Descartes to undertake studies in theoretical mathematics. Descartes responded by laying the foundations of analytic geometry, in which the properties of geometrical surfaces are expressed by algebraic equations.

In November 1619, after a period of particularly intense intellectual effort, Descartes experienced three dreams, the interpretation of which greatly influenced his life. He believed that he had been called by the Spirit of Truth to reconstruct human knowledge in such a way that it should embody the certainty heretofore possessed only by mathematics.

Descartes established residence in Holland in 1628, and remained there, except for brief visits to France, until 1649. He prepared a treatise—*Le Monde*—which set forth a mechanistic interpretation of the universe within which all change is caused by impact or pressure. He withheld the manuscript, however, upon learning of Galileo's condemnation by the Inquisition. He decided to prepare the ground for acceptance of *Le Monde* through other publications. Among these were the *Discourse on Method* (1637), to which were appended treatises on geometry, optics, and meteorology, as examples of application of the method, *Meditations on First Philosophy* (1641), and *Principles of Philosophy* (1644). *Le Monde* itself was published posthumously in 1664.

In 1649 Descartes accepted an invitation to become philosopher-in-residence to Queen Christina of Sweden. He died the following year.

Inversion of Francis Bacon's Theory of Procedure

Descartes agreed with Francis Bacon that the highest achievement of science is a pyramid of propositions, with the most general principles at the apex. But whereas Bacon sought to discover general laws by progressive inductive ascent from less general relations, Descartes sought to begin at the apex and work as far downwards as possible by a deductive procedure. Descartes, unlike Bacon, was committed to the Archimedean ideal of a deductive hierarchy of propositions.

Descartes demanded certainty for the general principles at the apex of the pyramid. In the service of this demand for certainty, he undertook systematically to doubt all judgements which he previously had believed to be true, in order to see if any of these judgements were beyond doubt. He concluded that certain of his judgements were indeed beyond doubt—that in so far as he thinks, he must exist, and that there must exist a Perfect Being.

Descartes reasoned that a Perfect Being would not create man in such a way that his senses and reason should systematically deceive him. Thus there must exist a universe external to the thinking self, a universe not opaque to man's cognitive faculties. Indeed, Descartes went further than this, claiming that any idea which is both clearly and distinctly present to the mind must be true.

According to Descartes, the clear is that which is immediately present to the mind. The distinct, on the other hand, is that which is both clear and unconditioned. The distinct is known *per se*; its self-evidence is independent of any limiting conditions. For instance, I may have a clear idea of the "bentness" of a stick partially immersed in water, without understanding the factors responsible for the appearance of "bentness". But to achieve a distinct idea of the "bentness" of the stick, I would have to understand the law of refraction and the way it applies to this particular case.

Primary Qualities and Secondary Qualities

After having established his own existence as a thinking being, and the existence of a benevolent God who guarantees that what is clearly and distinctly present to the mind is true, Descartes turned his attention to the created universe. He sought to discover that which is clear and distinct about physical objects. Commenting on the melting of a lump of wax, he declared that

while I speak and approach the fire what remained of the taste is exhaled, the smell evaporates, the colour alters, the figure is destroyed, the size increases, it becomes liquid, it heats, scarcely can one handle it, and when one strikes it, no sound is emitted. Does the same wax remain after this change? We must confess that it remains; none would judge otherwise. What then did I know so distinctly in this piece of wax? It could certainly be nothing of all that the senses brought to my notice, since all these things which fall under taste, smell, sight, touch, and hearing, are found to be changed, and yet the same wax remains ... abstracting from all that does not belong to the wax, let us see what remains. Certainly nothing remains excepting a certain extended thing which is flexible and movable.[1]

But how do we come to know this "extension" that constitutes the essence of the piece of wax? Descartes held that our knowledge of extension—the "real nature" of the wax—is an intuition of the mind. And this intuition of the mind is to be distinguished from the sequence of appearances that the wax presents to our senses. Descartes, like Galileo, distinguished between those "primary qualities" that all bodies must possess in order to be bodies, and the "secondary qualities"—colours, sounds, tastes, odours—that exist only in the perceptual experience of the subject.

Descartes reasoned that, since extension is the single property of bodies of which we have a clear and distinct idea, to be a body is to be extended. No vacuum can exist. Descartes took 'extension' to mean 'being filled by matter', and concluded that the concept "extension devoid of all matter" is a contradiction.[2]

But although he denied that a vacuum can exist in nature, Descartes did affirm certain of the methodological implications of classical atomism. He sought to interpret macroscopic processes in terms of submacroscopic interactions. An example is his interpretation of magnetic attraction. Descartes attributed the attraction of a magnet for a piece of iron to the emission from the magnet of invisible screw-shaped particles which pass through screwed channels present in the iron, thereby causing it to move. In addition, Descartes affirmed the atomist ideal of accounting for qualitative changes at the macroscopic level in terms of purely quantitative changes at the submacroscopic level. He restricted the subject-matter of science to those qualities that may be expressed in mathematical form and compared as ratios.

Descartes's vision of science thus combined the Archimedean, the Pythagorean, and the atomist points of view. For Descartes, the ideal of science is a deductive hierarchy of propositions, the descriptive terms of which refer to the strictly quantifiable aspects of reality, often at a submacroscopic level. No doubt he was influenced to accept this ideal by his early success in formulating analytic geometry. Descartes called for a universal mathematics to unlock the secrets of the universe, much as his analytic geometry had reduced the properties of geometrical surfaces to algebriac equations.

Unfortunately for this programme, Descartes also used the term 'extension' in a second sense. In order to describe the motions of bodies, he referred to bodies as occupying first one space and then another. For instance, if bodies A and B are bounded successively by bodies C and D, Descartes would speak of B as having moved into the "space" vacated by A.

But this "space" or "piece of extension", is not identical to any specific body. "Space", in this sense, is a relationship which a body has to other bodies. This dual usage of 'extension' is a serious equivocation. By Descartes's own standards, one must judge that he did not achieve a clear and distinct idea of "extension", his fundamental category for the interpretation of the universe.

The General Scientific Laws

Be that as it may, Descartes proceeded to derive several important physical principles from his understanding of extension. Buchdahl has pointed out that Descartes seemed to believe that because the concepts extension and motion are clear and distinct, certain generalizations about these concepts are *a priori* truths.[3] One such generalization is that all motion is caused by impact or pressure. Descartes maintained that, since no vacuum can exist, a given body is continually in contact with other bodies. It seemed to him that the only way a body can be moved is if the adjacent bodies on one side exert a greater pressure than the adjacent bodies on the other side. By restricting the causes of motion to impact and pressure, he denied the possibility of action-at-a-distance. Descartes defended a thoroughly mechanistic view of causation.

Descartes's Mechanistic Philosophy was a revolutionary doctrine in the seventeenth century. Many thinkers who accepted it believed it to be more scientific than rival views which entertained such "occult" qualities as magnetic forces and gravitational forces. From the Cartesian standpoint, to say that a body moved towards a magnet because of some force exerted by the magnet is to explain nothing. One might as well say that the body moved towards the magnet because it desired to embrace it.

Another important physical principle derived from the idea of extension is that all motion is a cyclical rearrangement of bodies. Descartes reasoned that, if one body changes its "location", a simultaneous displacement of other bodies is necessary to prevent a vacuum. Moreover, it is only by moving along a closed loop that a finite number of bodies can alter their positions without creating a vacuum.

Descartes maintained that God is the ultimate cause of motion in the universe. He believed that a Perfect Being would create a universe "all at once".* Descartes concluded that, since the matter of the universe was set in motion all at once, a Perfect Being would ensure that this motion be conserved perpetually. Otherwise, the universe would resemble a clock that eventually runs down, the product of an all-too-human workman.

From this most general principle of motion, Descartes derived three other laws of motion:

Law I. Bodies at rest remain at rest, and bodies in motion remain in motion, unless acted upon by some other body.

Law II. Inertial motion is straight-line motion.†

Law III (A). If a moving body collides with a second body, which second body has a greater resistance to motion than the first body has force to continue its own motion, then the first body changes its direction without losing any of its motion.

Law III (B). If the first body has greater force than the second body has resistance, then the first body carries with it the second, losing as much of its motion as it gives up to the second.

Descartes next deduced from these three laws seven rules of impact for specific kinds of collisions. These rules are incorrect, largely because Descartes took size, rather than weight, to be the determining factor in collisions. Of these rules of impact, the fourth is perhaps the most notorious. It states that, regardless of its speed, a moving body cannot budge a stationary body of greater size. In stating what he believed to be implied by the concepts "extension" and "motion," Descartes formulated a set of rules which are at variance with the observed motions of bodies.

Descartes claimed that the scientific laws he had elaborated were deductive consequences of his philosophical principles. In the *Discourse on Method* he wrote that

---

* Descartes did not explain why a Perfect Being, of necessity, would opt for a single act of creation, rather than a continuing creation of matter and motion.

† And not, as Galileo had held, circular motion.

I have first tried to discover generally the principles or first causes of everything that is or that can be in the world, without considering anything that might accomplish this end but God Himself who has created the world, or deriving them from any source excepting from certain germs of truths which are naturally existent in our souls.[4]

Much of the appeal of the Cartesian philosophy derives from its breadth of scope. Beginning with theistic-creationist metaphysical principles, Descartes proceeded to derive the general laws of the universe. Descartes's version of the pyramid of scientific truths is depicted at the top of the following page.

Empirical Emphases in Descartes's Philosophy of Science

The Limitations of A Priori Deduction

Descartes realized that one could proceed by deduction only a short distance from the apex of the pyramid. Deduction from intuitively self-evident principles is of limited usefulness in science. It can yield only the most general of laws. Moreover, since the fundamental laws of motion only place limits on what can happen under certain types of circumstances, innumerable sequences of events are consistent with these laws. Broadly speaking, the universe we know is but one of the indefinitely many universes that could have been created in accordance with these laws.

Descartes pointed out that one cannot determine, from mere consideration of the general laws, the course of physical processes. The law of conservation of motion, for instance, stipulates that, whatever process is considered, no loss of motion is incurred. But just how motion is redistributed among the bodies involved must be determined for each type of process. In order to deduce a statement about a particular effect, it is necessary to include among the premises information about the circumstances under which the effect occurred. In the case of the explanation of a physiological process, for example, the premises must include specific information about anatomical structure, in addition to the general laws of motion. Thus one important role for observation and experiment in Descartes's theory of scientific method is to provide knowledge of the conditions under which events of a given type take place.

It is at this point that the Baconian programme of compiling

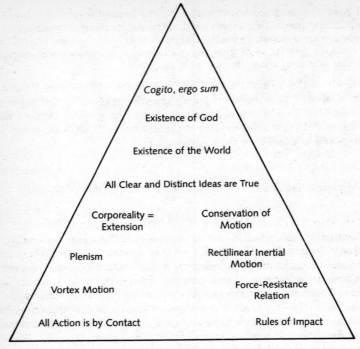

Descartes's Pyramid

natural histories and seeking correlations among phenomena is of value. Descartes conceded this much to Baconian science. He denied, however, that it is possible to establish important laws of nature by the collation and comparison of observed instances.

Role of Hypotheses in Science

A second important role of observation and experiment in Descartes's theory of scientific method is to suggest hypotheses which specify mechanisms that are consistent with the fundamental laws. Descartes held that a hypothesis is justified by its ability, in conjunction with the fundamental laws, to explain phenomena. The hypothesis must be consistent with the fundamental laws, but its specific content is to be adjusted to permit deduction of statements about the phenomena in question.

Frequently, Descartes suggested hypotheses that were based on analogies drawn from everyday experiences. He likened the motions of the planets to the revolution of bits of cork caught up in a whirlpool, the reflection of light to the bouncing of tennis balls on hard surfaces, and the action of the heart to the generation of heat in hay-mows. In each case the analogy to everyday experience was of crucial importance in the resultant theory.

It is quite likely that the use of pictorial analogies of this type contributed to the popularity of his theory of the universe. But more often than not, reliance on such analogies led Descartes astray.

A case in point is his explanation of the circulation of the blood. Descartes committed himself to an inappropriate analogy, and he ignored experimental evidence that counted against the analogy. According to Descartes, the heart, which generates heat on the model of spontaneous generation in hay-mows, vaporizes venous blood as it enters, thereby expanding the heart and propelling blood into the arterial system. Descartes's account conflicts with the facts. William Harvey had shown experimentally that the pulse of the blood into the arteries is accompanied by a *contraction* of the heart. Descartes had read Harvey's book on circulation, and had praised it, but elected to defend his own hypothesis nevertheless.[5]

Experimental Confirmation

It is on the issue of experimental confirmation that Descartes's theory of scientific method is most vulnerable. Clearly, he paid lip service, at least, to the value of experimental confirmation. He recognized, for instance, that a statement about a type of phenomena may be deduced from more than one set of explanatory premisses e.g.:

$$\frac{\text{laws of nature}}{\text{statement of relevant circumstances}}$$
$$\frac{\text{hypothesis 1}}{\therefore E}$$

$$\frac{\begin{array}{c} \text{laws of nature} \\ \text{statement of relevant circumstances} \\ \text{hypothesis 2} \end{array}}{\therefore E}$$

In such cases, Descartes specified that *other* effects be sought, such as are deducible from premises that include hypothesis 1, but are not deducible from premises that include hypothesis 2 (or vice versa).

However, Descartes's practice often did not match the sophistication of his writings about method. In general, he tended to regard experimentation as an aid in formulating explanations rather than as the touchstone of adequacy of such explanations.

Despite the fact that Descartes's interpretations often failed to fit facts, his theory of the universe possessed great appeal. It accorded due weight both to a desire for certainty and to an awareness of the complexity of phenomena. The general laws of nature supposedly were deductive consequences of necessary truths which must be acknowledged by any reflective individual.* And if 'quantity of motion' is interpreted as 'momentum', as Malebranche insisted, the resulting rules of impact do not conflict with experience. But these general laws explain phenomena only in conjunction with specific factual information, and often, hypotheses. It was possible to remove discrepancies between theory and observation by altering the associated hypotheses, thus leaving intact the general laws of nature. The existence of this flexibility within the Cartesian system was one reason for its continuing popularity (suitably modified) during the seventeenth and eighteenth centuries.

---

* Descartes was careful to emphasize that it was not necessary that God create the universe in accordance with the laws of the pyramid. The laws are not a constraint on God's creative activity. Indeed, Descartes held that it is within God's power to have created a world in which contradictions are realized. For instance, God could have created a world in which a circle has radii of different lengths, and in which mountains are present without valleys.[6] Needless to say, this possibility is beyond *human* understanding.

Nevertheless, Descartes consistently maintained that the essence of natural phenomena is extension and motion. And he often spoke as if the fundamental laws of motion—for this world that God *did* create—could not be other than they are. These laws are not mere empirical generalizations about what has been observed. Rather, they state clearly and distinctly comprehended insights into the structure of the universe.

Notes

[1] René Descartes, *Meditations on First Philosophy*, in *The Philosophical Works of Descartes*, trans. and ed. E. S. Haldane and G. R. T. Ross (New York: Dover Publications, 1955), i. 154.

[2] Descartes, *The Principles of Philosophy*, Haldane and Ross, i. 260–3.

[3] Gerd Buchdahl, *Metaphysics and the Philosophy of Science* (Oxford: Blackwell, 1969), 125.

[4] Descartes, *Discourse on the Method of Rightly Conducting the Reason*, Haldane and Ross, i. 121.

[5] Ibid. i. 112.

[6] Descartes, 'Letter to Mersenne (May 27, 1630)', 'Letter for Arnauld (July 29, 1648)', in *Descartes—Philosophical Letters*, trans. and ed. A. Kenny (Oxford: Clarendon Press, 1970), 15, 236–7.

8

Newton's Axiomatic Method

ISAAC NEWTON (1642–1727) was born in Woolsthorpe (Lincolnshire). His yeoman father died before Isaac's birth. Newton's mother remarried when he was three, and his upbringing was relegated largely to a grandmother, until the death of his stepfather in 1653.

Newton attended Trinity College, Cambridge, and received a BA degree in 1665. During 1665–7, Newton stayed at Woolsthorpe to avoid the plague. This was a period of immense creativity, in which Newton formulated the binomial theorem, developed the "method of fluxions" (calculus), constructed the first reflecting telescope, and came to realize the *universal* nature of gravitational attraction.

Newton was appointed Professor of Mathematics at Cambridge in 1669, and was elected a fellow of the Royal Society in 1672. Shortly thereafter, he communicated to the Society his findings on the refractive properties of light. An extended debate ensued with Robert Hooke and others. The controversy with Hooke deepened upon publication of the *Mathematical Principles of Natural Philosophy* (1687). Hooke complained that Newton had appropriated his position that planetary motions could be explained by a rectilinear inertial principle in combination with a $1/r^2$ force emanating from the sun. Newton replied that he had come to this conclusion before Hooke, and that only he could prove that a $1/r^2$ force law leads to elliptical planetary orbits.

Newton became Warden of the Mint in 1696 and displayed considerable talent for administration. He was elected President of the Royal Society in 1703, and from this vantage-point carried on a running feud with Leibniz over priorities in the development of the calculus. In 1704, Newton published the *Opticks*, a model of experimental inquiry. He included in the "Queries" at the end of this book a statement of his view of scientific method.

Throughout his life Newton studied the Biblical records from the standpoint of a Unitarian commitment. Extensive notes on the chronology of ancient kingdoms and the exegesis of *Daniel* have been found among his papers.

The Method of Analysis and Synthesis

Newton's comments about scientific method were directed primarily against Descartes and his followers. Descartes had sought to derive basic physical laws from metaphysical principles. Newton opposed this method of theorizing about nature. He insisted that the natural philosopher base his generalizations on a careful examination of phenomena. Newton declared that "although the arguing from Experiments and Observations by Induction be no Demonstration of general Conclusions, yet it is the best way of arguing which the Nature of Things admits of".[1]

Newton opposed the Cartesian method by affirming Aristotle's theory of scientific procedure. He referred to this inductive–deductive procedure as the "Method of Analysis and Synthesis". By insisting that scientific procedure should include both an inductive stage and a deductive stage, Newton affirmed a position that had been defended by Grosseteste and Roger Bacon in the thirteenth century, as well as by Galileo and Francis Bacon at the beginning of the seventeenth century.

Newton's discussion of the inductive–deductive procedure was superior to that of his predecessors in two respects. He consistently stressed the need of experimental confirmation of the consequences deduced by Synthesis, and he emphasized the value of deducing consequences that go beyond the original inductive evidence.

Newton's application of the Method of Analysis and Synthesis reached fruition in the investigations of the *Opticks*. For example, in a deservedly famous experiment, Newton passed a ray of sunlight through a prism such that an elongated spectrum of colour was produced on the far wall of a darkened room.

Newton's One-Prism Experiment

Newton applied the Method of Analysis to induce the explanatory principle that sunlight comprises rays of differing colours, and that each colour is refracted by the prism through a characteristic angle. This was not a simple inductive generalization on Newton's part. Newton did not affirm merely that all prisms under similar circumstances would produce spectra similar to those he had observed. His more important conclusion was about the nature of light itself, and it required an "inductive leap" to conclude that sunlight is made up of rays which have different refractive properties. After all, other interpretations of the evidence are possible. Newton might have concluded, for instance, that sunlight is indivisible, and that the spectral colours are produced instead by some sort of secondary radiation within the prism.

Given the "theory" that sunlight does comprise rays of different colours and refractive properties, Newton then applied the Method of Synthesis to deduce certain further consequences of the theory. He noted that if his theory were correct, then passing light of a particular colour through a prism should result in a deflection of the beam through the angle characteristic of that colour, but no resolution of the beam into other colours. Newton confirmed this consequence of his theory of colours by passing light from one small band of the spectrum through a second prism.[2]

Inductive Generalization and the Laws of Motion

Newton also claimed to have followed the Method of Analysis and Synthesis in his great work on dynamics, the *Mathematical*

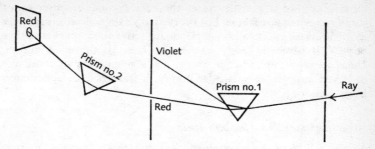

Newton's Two-Prism Experiment

Principles of Natural Philosophy (1686). In this volume, he reported that he had formulated the three laws of motion upon application of the Method of Analysis. Newton declared that in experimental philosophy "particular propositions are inferred from the phenomena, and afterwards rendered general by induction. Thus it was that the impenetrability, the mobility, and the impulsive force of bodies, and the laws of motion and of gravitation, were discovered."[3]

Newton did not discuss the nature of the inductive process which proceeds from phenomena to particular propositions to the laws of motion. Whether or not it is correct to say that the laws of motion were discovered upon application of the Method of Analysis depends on how broadly one construes "induction".

Aristotle, for instance, admitted intuitive insight as a bona fide inductive method. Aristotle's theory of procedure thus could account for generalizations about weightless, infinitely rigid levers, ideal pendulums, and inertial motion. Indeed, it would be difficult to find a scientific interpretation whose origin could not be attributed to intuitive insight.

Most natural philosophers, however, have taken a more restricted view of induction, limiting it to a small number of techniques for generalizing the results of observation. These techniques include simple enumeration, and the methods of agreement and difference.

It is clear that Newton's Laws were not discovered upon application of these inductive techniques. Consider the first law. It specifies the behaviour of those bodies which are under the influence of no impressed forces. But no such bodies exist. And even if such a body did exist, we could have no knowledge of it.

Observation of a body requires the presence of an observer or some recording apparatus. But on Newton's own view, every body in the universe exerts a gravitational attractive force on every other body. An observed body cannot be free of impressed forces. Consequently, the law of inertia is not a generalization about the observed motions of particular bodies. It is, rather, an abstraction from such motions.

Absolute Space and Absolute Time

Moreover, Newton maintained that the three laws of motion specify how bodies move in Absolute Space and Absolute Time. This is a further abstraction on Newton's part. Newton contrasted Absolute Space and Time with their "sensible measures" which are determined experimentally.

Newton's distinction between the "true motions" of bodies in Absolute Space and Time and the "sensible measures" of these motions has a Platonic ring that suggests a dichotomy of reality and appearance. On Newton's view, Absolute Space and Absolute Time are ontologically prior to individual substances and their interactions. He believed, moreover, that an understanding of sensible motions can be achieved in terms of true motions in Absolute Space.

Newton recognized that to establish that a sensible measure of a body's motion is its true motion, or that a sensible motion is related in some specific way to its true motion, it would be necessary to specify both Absolute temporal intervals and co-ordinates in Absolute Space. But he was not certain that these requirements can be met.

With respect to Absolute Time, Newton declared that "it may be, that there is no such thing as an equable motion, whereby time may be accurately measured. All motions may be accelerated and retarded, but the flowing of absolute time is not liable to any change."[4] However, Newton did indicate that some sensible measures of time are preferable to others. He suggested that for the definition of temporal intervals, the eclipses of Jupiter's moons and the vibrations of pendulums are superior to the apparent motion of the sun around the Earth.[5]

But even if Absolute Time could be measured, it still would be necessary to locate a body in Absolute Space before its absolute motion could be determined. Newton was convinced that Absolute

Space must exist, and he advanced both theological arguments and physical arguments for its existence, but he was less certain that bodies could be located in this space.

Newton maintained on theological grounds that since the universe was created *ex nihilo*, there must exist a receptacle within which created matter is distributed. He suggested that Absolute Space is an "emanent effect" of the Creator, a "disposition of all being" which is neither an attribute of God nor a substance coeternal with God. Newton criticized Descartes's identification of extension and body as offering a path to atheism, since, according to Descartes, we can achieve a clear and distinct idea of extension independently of its nature as a creation of God.[6]

The most important of Newton's physical arguments for the existence of Absolute Space was his analysis of the motion of a rotating, water-filled bucket.* He noted that if such a bucket were suspended from a twisted rope and allowed to rotate as the rope unwinds, the water surface remains a plane for a time and only gradually assumes a concave shape. At length the water rotates at the same rate as the bucket. Newton's experiment showed that the deformation of the water surface could not be correlated with an acceleration of the water relative to the bucket, since the water surface is successively a plane and concave when there is a relative acceleration, and since the water surface may be either a plane or concave when there is no relative acceleration.

Newtons's Bucket Experiment

Event	Acceleration of water relative to bucket in earth-centred co-ordinate system	Surface of water
1. Bucket stationary	no	plane
2. Bucket released	yes	plane
3. At maximum rotation	no	concave
4. Bucket arrested	yes	concave
5. Water at rest	no	plane

* Many interpreters have taken Newton to have cited the bucket experiment as evidence for the existence of Absolute Space. Ronald Laymon has argued, however, that Newton described the rotating bucket merely to illustrate that absolute motions can be distinguished from relative motions on the prior assumption that Absolute Space does exist.[7]

Newton maintained that deformation of the water surface indicates that a force is acting. And the second law of motion associates force and acceleration. But this acceleration of the water is an acceleration with respect to what? Newton concluded that since the acceleration associated with deformation is not an acceleration relative to the bucket, it must be an acceleration with respect to Absolute Space.[8]

Subsequently, numerous writers have pointed out that Newton's conclusion does not follow from his experimental findings. Ernest Mach, for example, suggested that the deformation be correlated, not with an acceleration with respect to Absolute Space, but with an acceleration with respect to the fixed stars.[9]

However, even if Newton were correct to conclude that the bucket experiment demonstrates the existence of an absolute motion, this would not suffice to specify a system of co-ordinates for locating positions in Absolute Space. Newton conceded this. Moreover, he admitted that there may be no single body which is at rest with respect to Absolute Space, and which may serve as a reference point for measuring distances in this space.[10]

Newton thus admitted that it may not be possible to achieve a wholly satisfactory correspondence between observed motions and true motions in Absolute Space. His explicit discussion of this problem of correspondence indicates that he followed an axiomatic method in the *Principia* rather than the inductive method of Analysis.

An Axiomatic Method

There are three stages in Newton's axiomatic method. The first stage is the formulation of an axiom system. On Newton's view, an axiom system is a deductively organized group of axioms, definitions, and theorems. Axioms are propositions that cannot be deduced from other propositions within the system, and theorems are the deductive consequences of these axioms. The three laws of motion are the axioms of Newton's theory of mechanics. They stipulate invariant relations among such terms as 'uniform motion in a right line', 'change of motion', 'impressed force', 'action', and 'reaction'. The axioms are:

I. Every body continues in its state of rest, or of uniform motion in a right line, unless it is compelled to change that state by forces impressed upon it.

II. The change of motion is proportional to the motive force impressed; and is made in the direction of the right line in which that force is impressed.

III. To every action there is always opposed an equal reaction: or, the mutual actions of two bodies upon each other are always equal, and directed to contrary parts.[11]

Newton clearly distinguished the "absolute magnitudes" which appear in the axioms from their "sensible measures" which are determined experimentally. The axioms are *mathematical principles* of natural philosophy which describe the true motions of bodies in Absolute Space.

The second stage of the axiomatic method is to specify a procedure for correlating theorems of the axiom system with observations. Newton usually required that axiom systems be linked to events in the physical world.

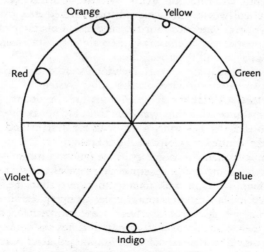

Newton's Theory of Colour-Mixing

However, he did submit for consideration a Theory of Colour-Mixing in which the axiom system was not properly linked to experience.[12] Newton specified that a circle be drawn and be subdivided into seven wedges—one for each of the "principal colours" of the spectrum—such that the widths of the wedges are proportional to the musical intervals in the octave. He further specified that the "number of rays" of each colour in the mixture be represented by a circle of greater or smaller radius located at the midpoint of the arc for each colour present in the mixture. Newton indicated that the centre of gravity of these circles gives the resultant colour of the mixture.

Newton's axiom on slicing the pie to satisfy musical harmonies is reminiscent of Kepler's Pythagorean speculations. The axiom certainly is not an inductive generalization. Nevertheless, even though there is no evidence in support of the pie-slicing axiom, the theory would be useful if the results of mixing colours could be calculated from it. But Newton failed to provide an empirical interpretation for the phrase 'number of rays'. Since he did not stipulate how the diameters of the circles are to be determined, Newton's theory of colour-mixing has no empirical significance.

Newton's mechanics, on the other hand, does have empirical significance. He did link his axiom system for mechanics to events in the physical world. He achieved the required link by selecting "Rules of Correspondence" for the conversion of statements about Absolute spatial and temporal intervals into statements about measured spatial and temporal intervals.

In the case of spatial intervals, Newton asserted as a "hypothesis" that the centre of gravity of the solar system is immovable, and therefore a suitable reference point for the determination of Absolute distances. He thus was able to apply his axiom system to actual motions by selecting a co-ordinate system the origin of which is the centre of gravity of the solar system.

I. Bernard Cohen has suggested that Newton meant by "hypothesis" in this context a proposition that he was unable to prove.[13] But although Newton was unable to prove that the centre of gravity of the solar system is immovable, his hypothesis is consistent with his interpretation of the bucket experiment. On this inter-pretation, the recession of water towards the walls of the bucket is an acceleration with respect to Absolute Space. According to Newton, this centrifugal acceleration typifies those effects which distinguish motions with respect to Absolute Space from merely relative motions.[14] Newton believed that "the motion which causes the Earth to endeavour to recede from the Sun" is likewise an Absolute Motion.[15] Since the centre of gravity of the solar system is the "centre" of this motion of revolution (at least in so far as the motion is approximately circular), Newton's hypothesis fits in with his views on Absolute Motion.

In the case of temporal intervals, Newton did not specify that any one periodic process should be taken as the measure of

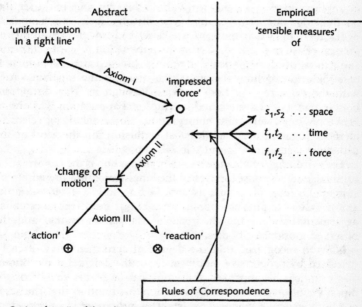

1. Centre of gravity of the solar system taken as the centre of Absolute Space.
2. Selection of the 'best measure' of Absolute Time.
3. Moving bodies construed as systems of indefinitely large numbers of point-masses.
4. Specification of experimental procedures to measure values of impressed forces.

Newton's Interpreted Axiom System for Mechanics

Absolute Time. However, by reading between the lines, one can interpret Newton to have suggested a procedure to link Absolute Time with its sensible measures. Such a link might be established by examining time-dependent sequences which have been determined using various different methods of measuring time. For example, if the distance–time relationship for balls rolled down inclined planes is "more regular" when time is measured by the swings of a pendulum than when time is measured by the weight of water flowing through a hole in a pail, then the pendulum clock is the better "sensible measure" of Absolute Time.[16]

Newton thus carefully distinguished the abstract status of an axiom system from its application to experience. See the diagram on the previous page. This distinction may be illustrated as follows:

Newton enforced the distinction between an axiom system and its application to experience throughout the *Principia*. In the section on fluid dynamics, for example, he distinguished "mathematical dynamics", in which motions are described under various hypothetical resistive conditions, from its application to experience. An application of mathematical dynamics is achieved after experimental determination of how the resistance of a specific medium varies with the velocity of a body moving through it. This distinction between an axiom system and its empirical application was one of Newton's most important contributions to the theory of scientific method. It raised to a new level of sophistication the ideal of the deductive systematization of scientific knowledge.

The third stage of Newton's axiomatic method is the confirmation of the deductive consequences of the empirically interpreted axiom system. Once a procedure is specified to link the terms of the axiom system to phenomena, the investigator must seek to establish agreement between the theorems of the axiom system and the observed motions of bodies.

Newton recognized that the degree of agreement may often be increased by progressive modification of the original assumptions. For instance, he improved the empirical fit of his theory of the moon's motion by modifying the initial assumption that the earth is a homogeneous sphere. This feedback procedure is an important aspect of what I. B. Cohen has termed the "Newtonian Style" in natural philosophy.[17]

Newton himself established extensive agreement between his empirically interpreted axiom system for mechanics and the motions of celestial and terrestrial bodies. An illustration is his experiments with colliding pendulums. Newton showed that after appropriate corrections are made for air resistance, action and reaction are equal regardless of whether the pendulum bobs are composed of steel, glass, cork, or wool.

Newton thus affirmed and practised *two* theories of scientific procedure—the Method of Analysis and Synthesis, and an Axiomatic Method. I think that it does not detract from Newton's genius to point out that he did not keep in mind consistently the distinction between these two theories of procedure.

The Method of Analysis and Synthesis and the Axiomatic Method share as a common objective the explanation and prediction of phenomena. But they differ in an important respect, particularly if one takes a narrow view of what techniques qualify as "induction". The natural philosopher who follows the Method of Analysis seeks to generalize from the results of observation and experiment. The Axiomatic Method, by contrast, places greater emphasis on the creative imagination. The natural philosopher who adopts this method may begin anywhere. But the axiom system he creates is relevant to science only if it can be linked to what can be observed.

"Hypotheses Non Fingo"

Newton agreed with Galileo that primary qualities are the proper subject-matter of physics. According to Newton, the starting-point and end-point of scientific inquiry is the determination of the values of "manifest qualities", those aspects of phenomena that may be measured experimentally.

Newton sought to restrict the content of his "experimental philosophy" to statements about manifest qualities, "theories" derived from these statements, and queries directive of further inquiry. In particular, he sought to exclude "hypotheses" from experimental philosophy.

Newton's use of the terms 'theory' and 'hypothesis' does not conform to modern usage. He applied the term 'theory' to invariant relations among terms designating manifest qualities. He sometimes spoke of these invariant relations as relations "deduced from" phenomena, but he most likely meant by this that there was very strong inductive evidence for certain of these relations.

'Hypotheses', in one of Newton's usages,* are statements about terms that designate "occult qualities" for which no measuring procedures are known.

Newton was quick to take offence whenever his experimentally based "theories" were labelled "hypotheses". For example, when the mathematician Pardies incautiously referred to Newton's theory of colours as a "very ingenious hypothesis",[18] Newton promptly corrected him. Newton emphasized that there was conclusive experimental evidence that sunlight comprises rays of differing colours and refractive properties. He distinguished carefully his "theory" that light has certain properties of refraction, from any "hypothesis" about waves or corpuscles by which these properties might be explained.[19]

Newton defended a similar position on the "theory" of gravitational attraction. He insisted that he had established the existence of gravitational attraction and its mode of operation, thereby accounting for the motions of the planets, the tides, and diverse other phenomena. But he did not wish to jeopardize this "theory" by tying it to a particular hypothesis about the underlying cause of the attraction. "I feign no hypotheses", he wrote.[20]

His injunction was directed primarily against "explanations" of gravitational attraction in terms of the Cartesian hypothesis of invisible swirling vortices of ether. Newton demonstrated in the *Principia* that Descartes's Vortex Hypothesis had consequences that are not in agreement with the observed motions of the planets.

Yet in other contexts, Newton was willing to entertain hypotheses that explain correlations among manifest qualities. Indeed, he himself flirted with a hypothesis about an ethereal medium which produces gravitational attraction. However, Newton emphasized that the function of such hypotheses is to direct future research, and not to serve as premises for sterile disputation.

The Rules of Reasoning in Philosophy

To direct the search for *fruitful* explanatory hypotheses, Newton suggested four regulative principles, referred to as "hypotheses"

---

* I. B. Cohen has discussed nine meanings of 'hypothesis' in Newton's writings (*Franklin and Newton*, 138–40).

in the first edition of the *Principia*, and "rules of reasoning in philosophy" in the second edition. These regulative principles are:

I. We are to admit no more causes of natural things than such as are both true and sufficient to explain their appearances.

II. Therefore to the same natural effects we must, as far as possible, assign the same causes.

III. The qualities of bodies, which admit neither intensification nor remission of degrees, and which are found to belong to all bodies within the reach of our experiments, are to be esteemed the universal qualities of all bodies whatsoever.

IV. In experimental philosophy we are to look upon propositions inferred by general induction from phenomena as accurately or very nearly true, notwithstanding any contrary hypotheses that may be imagined, till such time as other phenomena occur, by which they may either be made more accurate, or liable to exceptions.[21]

In support of Rule I, Newton appealed to a principle of parsimony, declaring that nature "affects not the pomp of superfluous causes". But exactly what Newton meant, or should have meant, by a "true cause" has been a subject of some debate. For instance, both William Whewell and John Stuart Mill criticized Newton for failing to specify criteria for the identification of true causes. Whewell remarked that if Newton meant to restrict the "true cause" of a type of phenomena to causes already known to be effective in producing other types of phenomena, then Rule I would be overly restrictive. It would preclude the introduction of new causes. However, Whewell was not certain that this was Newton's intended meaning. He noted that Newton may have meant only to restrict the introduction of causes to those "similar in kind" to causes that previously have been established. Whewell observed that, thus interpreted, Rule I would be too vague to guide scientific inquiry. Any hypothetical cause could be claimed to display *some* similarity to previously established causes. Having dismissed these inadequate alternatives, Whewell suggested that what Newton should have meant by a "true cause" is a cause represented in a theory, which theory is supported by inductive evidence acquired from analysis of diverse types of phenomena.*

* Whewell's concept of a "consilience of inductions" is discussed in Chapter 9.

Mill likewise interpreted "true cause" so as to reflect his own philosophical position. Consistent with his view of induction as a theory of proof of causal connection, Mill maintained that what distinguishes a "true cause" is that its connection with the effect ascribed to it be susceptible to proof by independent evidence.[†]

Commenting on Rule III, Newton indicated that the qualities which satisfy the rule include extension, hardness, impenetrability, mobility, and inertia. Newton maintained that these qualities should be taken to be the universal qualities of all bodies whatsoever. Moreover, he insisted that these also are the qualities of the minute parts of bodies. In Query 31 of the *Opticks*, he set forth a research programme to uncover the forces that govern the interactions of the minute parts of bodies. Newton expressed the hope that the study of short-range forces would achieve an integration of physico-chemical phenomena such as changes of state, solution, and the formation of compounds, in much the same way as the principle of universal gravitation had achieved the integration of terrestrial and celestial dynamics. Subsequently, Newton's research programme received theoretical development from Boscovich and Mossotti, and practical implementation in the electromagnetic researches of Faraday and the various attempts to measure the elective affinities of the chemical elements.[‡]

The Contingent Nature of Scientific Laws

Newton repudiated the Cartesian programme of deducing scientific laws from indubitable metaphysical principles. And he denied that a necessary knowledge of scientific laws can be achieved in any manner. According to Newton, the natural philosopher may establish that phenomena are related in a certain way, but cannot establish that the relation could not be otherwise.

It is true that Newton did suggest that if we could know the forces that operate on the minute particles of matter, we could understand why macroscopic processes occur in the ways they do. But Newton did not maintain that such knowledge would constitute

---

[†] Mill's view of causal relation is discussed in Chapter 9.

[‡] The role of Newton's research programme in 18th-c. science has been discussed by A. Thackray in *Atoms and Powers* (Cambridge, Mass.: Harvard University Press, 1970).

a necessary knowledge of nature. On the contrary, he held that all interpretations of natural processes are contingent and subject to revision in the light of further evidence.

Notes

[1] Isaac Newton, *Opticks* (New York: Dover Publications, 1952), 404.

[2] Ibid. 45–8.

[3] Newton, *Mathematical Principles of Natural Philosophy*, trans. A. Motte, revised by F. Cajori (Berkeley, Calif.: University of California Press, 1962), ii. 547.

[4] Ibid. i. 8.

[5] Ibid. i. 7–8.

[6] Newton, *Unpublished Scientific Papers of Isaac Newton*, trans. and ed. A. R. Hall and M. B. Hall (Cambridge: Cambridge University Press, 1962), 132–43.

[7] Ronald Laymon, 'Newton's Bucket Experiment', *J. Hist. Phil.* 16 (1978), 399–413.

[8] Newton, *Mathematical Principles*, i. 10–11.

[9] Ernst Mach, *The Science of Mechanics*, trans. T. J. McCormack (La Salle, Ill.: Open Court Publishing Co., 1960), 271–97.

[10] Newton, *Mathematical Principles*, i. 8.

[11] Ibid. i. 13.

[12] Newton, *Opticks*, 154–8.

[13] I. Bernard Cohen, *Franklin and Newton* (Philadelphia: The American Philosophical Society, 1956), 139.

[14] Newton, *Mathematical Principles*, i. 10.

[15] Newton, *Unpublished Scientific Papers*, 127.

[16] e.g. see S. Toulmin, 'Newton on Absolute Space, Time, and Motion', *Phil. Rev.* 68 (1959); E. Nagel, *The Structure of Science* (New York: Harcourt, Brace, and World, 1961), 179–83.

[17] I. Bernard Cohen, *The Newtonian Revolution* (Cambridge: Cambridge University Press, 1980), 52–154.

[18] Ignatius Pardies, 'Some Animadversions on the Theory of Light of Mr. Isaac Newton', in *Isaac Newton's Papers and Letters on Natural Philosophy*, ed. I. B. Cohen (Cambridge, Mass.: Harvard University Press, 1958), 86.

[19] Newton, 'Answer to Pardies', in *Isaac Newton's Papers and Letters on Natural Philosophy*, 106.

[20] Newton, *Mathematical Principles*, ii. 547. See also A. Koyré, *Newtonian Studies* (Cambridge, Mass.: Harvard University Press, 1965), 35–6.

[21] Newton, *Mathematical Principles*, ii. 398–400.

9

Analyses of the Implications of the New Science for a Theory of Scientific Method

I. THE COGNITIVE STATUS OF SCIENTIFIC LAWS

JOHN LOCKE (1632–1704) was born at Wrington (Somerset). He was educated at Oxford and was appointed lecturer in Greek and philosophy there in 1660. Subsequently, he became interested in medicine and obtained a licence to practise, again from Oxford.

In 1666, Locke joined the service of the first Earl of Shaftesbury, and became physician, friend, and adviser to this influential politician. Upon Shaftesbury's fall from power, Locke chose exile in Holland. It was during his stay in Holland that Locke completed his *Essay Concerning Human Understanding* (1690), in which he set forth his views on the prospects and limitations of science. Locke's political fortunes improved upon the accession of William of Orange in 1689. He returned to England and accepted a position in the Civil Service.

GOTTFRIED WILHELM LEIBNIZ (1646–1716) was the son of the Professor of Moral Philosophy at the University of Leipzig. An omnivorous reader, Leibniz studied philosophy at his father's university, and jurisprudence at Jena.

Leibniz spent much of his adult life at court, first at Mainz and later at Hanover. During this service he was entrusted with diplomatic missions which enabled him to establish contacts with numerous political and intellectual leaders. Leibniz worked tirelessly for legal reform, for Protestant religious unification, and for the advancement of science and technology. He maintained extensive correspondences with the leading thinkers of his day and actively promoted scientific co-operation by means of his membership in the Royal Society, the French Academy, and the Prussian Academy. It is ironic that his later years were marked by bitter polemics with the followers of Newton over priorities in the invention of the calculus.

DAVID HUME (1711–76) enrolled to study law at the University of Edinburgh, but left without receiving a degree. He neglected his legal studies for the pursuit of philosophy. Hume spent several years at Rheims and La Flèche, where he completed work on the *Treatise of Human Nature* (1739–40).

Hume was greatly disappointed with the reception accorded this book which "fell deadborn from the press". Undaunted, he revised and popularized the *Treatise* in *An Enquiry Concerning Human Understanding* (1748). Hume also published an *Enquiry Concerning the Principles of Morals* (1751), and a lengthy *History of England* (1754–62).

Hume was rebuffed in his attempts to secure positions at the Universities of Edinburgh and Glasgow. His opponents alleged heresy and even atheism. In 1763 Hume was appointed secretary to the British ambassador to France, and subsequently was lionized by Parisian society.

IMMANUEL KANT (1724–1804) spent his entire life in the immediate vicinity of his native Königsberg. He studied philosophy and theology at the University of Königsberg, and became Professor of Logic and Metaphysics there in 1770. Kant's views on the importance of regulative principles in scientific inquiry are set forth in *Critique of Pure Reason* (1781), and *Critique of Judgement* (1790).

Locke on the Possibility of a Necessary Knowledge of Nature

John Locke, who like Newton was committed to atomism, specified the conditions that would have to be fulfilled to achieve a necessary knowledge of nature. According to Locke, we would have to know both the configurations and motions of atoms and the ways in which the motions of atoms produce ideas of primary and secondary qualities in the observer. He noted that if these two conditions could be fulfilled, then we would know *a priori* that gold must dissolve in *aqua regis* but not *aqua fortis*, that rhubarb must have a purgative effect, and that opium must make a man sleepy.[1]

Locke held that we are ignorant of the configurations and motions of atoms. But his usual position was that this ignorance is a contingent matter, a question of the extreme minuteness of atoms. In principle, we might be able to overcome this ignorance. But even if this were achieved, we still could not reach a necessary knowledge of phenomena. This is because we are ignorant of the ways in which atoms manifest certain powers. Locke held that the atomic constituents of a body possess the power, in virtue of their motions, to produce in us ideas of secondary qualities such as colours and sounds. Moreover, the atoms of a particular body have the power to affect the atoms of other bodies so as to alter the ways in which these bodies affect our senses.[2] At one point, Locke declared that only by divine revelation could we know the ways in which atomic motions produce these effects in us.[3]

In some passages, Locke held that an unbridgeable epistemological gap separates the "real world" of atoms and the realm of ideas that constitutes our experience. And he expressed no interest in entertaining hypotheses about atomic structure. It is a curious feature of Locke's philosophy of science that although he consistently attributed macroscopic effects to atomic interactions, he made no attempt to correlate specific effects with particular hypotheses about atomic motions. As Yolton has pointed out, Locke instead recommended for science a Baconian methodology of correlation and exclusion, based on the compilation of extensive natural histories.[4] This involved a shift in focus from "real essences"—the atomic configurations of bodies—to "nominal essences"—the observed properties and relations of bodies.

Locke insisted that the most that can be achieved in science is a collection of generalizations about the association and succession of "phenomena". These generalizations are probable at best, and do not satisfy the rationalist ideal of necessary truth. In this vein, Locke sometimes downgraded natural science. In one passage, he conceded that the trained scientist views nature in a more sophisticated way than does an untrained observer, but he insisted that this is "but. judgment and opinion, not knowledge and certainty".[5]

Yet in other passages, Locke drew back from the sceptical possibilities implicit in his distinction between the primary properties of the atomic constituents of bodies, which properties exist independently of our perceptual experiences, and our ideas of secondary qualities. He believed that there do exist necessary connections in nature, even though these connections are opaque to human understanding. Locke often used the term 'idea' in such a way as to bridge the epistemological gap. In this usage, "ideas" are effects of operations in the "real world" of atoms. The idea of a red patch, for example, is a possession of a perceiving subject, but it also is an effect somehow produced by processes external to the subject (in normal viewing situations at least). Locke was confident that it is the motions of the atomic constituents of matter that give rise to our ideas of colours and tastes, even though we cannot learn just how this takes place. It remained for Berkeley and Hume to demand that the warrant be produced for this assumption.

Leibniz on the Relationship between Science and Metaphysics

Locke's contemporary Leibniz gave a more optimistic assessment of what can be achieved in science. Leibniz was a practising scientist who made important contributions to mathematics and physics. And he confidently extrapolated from his scientific findings to metaphysical assertions. Indeed, Leibniz set up a two-way commerce between scientific theories and metaphysical principles. Not only did he support his metaphysical principles by analogical arguments based on scientific theories. He also employed metaphysical principles to direct the search for scientific laws.

A case in point is the relationship between studies of impact phenomena and the principle of continuity. Leibniz used the principle of continuity to criticize Descartes's rules of impact. He noted that, according to Descartes, if two bodies of equal size and speed collide head-on, their speeds after impact are the same, but in reversed directions; but that if one body is larger than the other, both bodies proceed after impact in the direction in which the larger body was travelling. Leibniz objected that it is unreasonable that an infinitesimal addition of matter would result in a discontinuous change of behaviour.[6] And having corrected Descartes's rules of impact, Leibniz was quite willing to appeal to impact phenomena to support the ontological claim that nature invariably acts so as to avoid discontinuities.

A similar reciprocal interaction is present in Leibniz's discussion of the relationship between *extremum* principles in physics and the principle of perfection. For instance, he argued that because nature always selects the easiest, or most direct, course of action from among a set of alternatives, the passage of a light ray from one medium into another obeys Snel's Law.[*] Leibniz derived Snel's Law by applying the differential calculus which he had developed to the condition that the "path difficulty" of the ray (the path length times the resistance of the medium) is a minimum. And he took his success in this enterprise as support for the metaphysical principle that God governs the universe in such a way that a maximum of "simplicity" and "perfection" be realized.[7]

Further evidence of Leibniz's view of the interdependence of physics and metaphysics is the relationship between the conservation of *vis viva* (mv^2) and the principle of monadic activity. On the one hand, Leibniz argued analogically from the conservation of *vis viva* in physical processes to a characterization of being-as-such as an "internal striving". On the other hand, his conviction that monadic activity on the metaphysical plane must have its correlate on the physical plane directed his attention to a search for some "entity" that is conserved in physical interactions.

Buchdahl has called attention to the importance of Leibniz's metaphysical commitment by contrasting the analyses of collision processes given by Huygens and Leibniz. Whereas Huygens merely noted in passing that mv^2, regarded as the product of mathematical

---

[*] Snel's law states that '$\dfrac{\sin i}{\sin r}$ = constant' for any pair of media, where i is the angle of incidence of a light ray, and r is its angle of refraction.

parameters, remained constant in such processes, Leibniz "substantialized" *vis viva* and held that its conservation was a general physical principle.[8]

Leibniz sought to interpret the universe in such a way that the mechanistic world-view, which focuses on material and efficient causation, is supported by teleological considerations. *Extremum* principles, conservation principles, and the principle of continuity were well suited to effect the desired integration of the mechanistic and teleological standpoints. In the case of *extremum* principles, for example, the teleological connotation is that natural processes occur in certain ways *in order that* certain quantities achieve a minimum (or maximum) value. It is a short step, and one that Leibniz was anxious to take, to the position that a Perfect Being created the universe in such a way that natural processes satisfy these principles.

Locke had bemoaned the fact that we cannot advance from a knowledge of the association of qualities to a knowledge of the internal consitutions or "real essences" of things. Leibniz took quite a different attitude towards this epistemological gap. He conceded that, at the level of phenomena, scientists can reach only probability, or "moral certainty". But he was convinced that the general metaphysical principles he had formulated were necessary truths. Of necessity, individual substances (monads) unfold in accordance with a principle of perfection that ensures their harmonious interrelation. And we can be certain that this monadic activity "underlies" phenomena. But we cannot know that the metaphysical principles *must* be instantiated, at the level of phenomena, in one particular way.

As a rule, Leibniz emphasized the certainty of his metaphysical principles rather than the contingent nature of empirical knowledge. His dominant posture was one of optimism. Indeed, at times he appeared to claim more than probability for empirical generalizations. This inconsistency perhaps may be attributed to an overriding concern to establish the dependence of the phenomenal realm on the metaphysical realm.

Leibniz recognized that a picture of a metaphysical realm "behind" phenomena is of interest only if there are strong links between the two realms. The strongest possible links would be deductive relationships between metaphysical principles and empirical laws. Given the necessary status of metaphysical principles, deductive relationships would extend the domain of necessary connectedness into the realm of phenomena.

Leibniz flirted with this possibility. He employed an analogy based on the theory of infinite series to suggest that there are strong links between the two realms. The analogy is that metaphysical principles are related to physical laws much as the law that generates an infinite series is related to the particular members of that series.[9]

But even if one were to accept the force of this analogy, this would not establish that metaphysical principles *imply* empirical laws. One cannot deduce, from the law of a series alone,

$$\left(\text{e.g.} \sum_{n=1}^{\infty} \frac{1}{n^2} \right)^*$$

the value of a particular member of the series. The position of the term in the series must be specified (e.g. $n = 5$). Similarly, one cannot deduce from metaphysical principles alone specific empirical laws. The way in which a metaphysical principle is realized in experience must be specified. But on Leibniz's own admission, we cannot know that a metaphysical principle *must* be realized in one specific way.

I think Leibniz was aware that the infinite-series analogy could not be pressed. On other occasions he spoke of physical forces as the "echoes" of metaphysical forces,[10] a characterization that is extremely vague. And to retreat to this position was to leave unresolved the general problem of the relationship between the two realms, as well as the particular problem about the cognitive status of *extremum* principles and conservation principles *as applied in science*.

Hume's Scepticism

David Hume extended and made consistent Locke's sceptical approach to the possibility of a necessary knowledge of nature. Hume consistently denied that a knowledge of atomic configurations

---

* $\sum_{n=1}^{\infty} \frac{1}{n^2} = 1 + \frac{1}{4} + \frac{1}{9} + \frac{1}{16} + \ldots = \frac{\pi^2}{6}.$

and interactions—even if it could be achieved—would constitute a necessary knowledge of nature. According to Hume, even if our faculties were "fitted to penetrate into the internal fabric" of bodies, we could gain no knowledge of a necessary connectedness among phenomena. The most we could hope to learn is that certain configurations and motions of atoms have been constantly conjoined with certain macroscopic effects. But knowing that a constant conjunction has been observed is not the same thing as knowing that a particular motion *must* produce a particular effect. Hume held that Locke was wrong to suggest that if we knew the atomic configuration of gold then we would understand without trial that this substance must be soluble in *aqua regia*.

Hume's denial of the possibility of a necessary knowledge of nature was based on three explicitly stated premises: (1) all knowledge may be subdivided into the mutually exclusive categories "relations of ideas" and "matters of fact"; (2) all knowledge of matters of fact is given in, and arises from, sense impressions; and (3) a necessary knowledge of nature would presuppose knowledge of the necessary connectedness of events. Hume's arguments in support of these premises were widely influential in the subsequent history of the philosophy of science.

Subdivision of Knowledge

Hume maintained that statements about relations of ideas and statements about matters of fact differ in two respects. The first respect is the type of truth-claim that can be made for the two types of statements. Certain statements about relations of ideas are necessary truths. For instance, given the axioms of Euclidean geometry, it could not be otherwise than that the sum of angles of a triangle is 180 degrees.* To affirm the axioms and deny the theorem is to construct a self-contradiction. Statements about matters of fact, on the other hand, are never more than contingently true. The denial of an empirical statement is not a self-contradiction; the state of affairs described could have been otherwise.

---

* Hume denied that the propositions of geometry were necessary truths in *A Treatise of Human Nature* (1739), but subsequently changed his mind. In the *Enquiry Concerning Human Understanding* (1748), he held that geometrical propositions, as well as the propositions of arithmetic and algebra, are necessary truths.

The second point of difference is the method followed to ascertain the truth or falsity of the respective types of statements. The truth or falsity of statements about relations of ideas is established independently of any appeal to empirical evidence. Hume subdivided statements about relations of ideas into those which are intuitively certain and those which are demonstratively certain. For example, the axioms of Euclidean geometry are intuitively certain; their truth is established upon examination of the meanings of their component terms. The Euclidean theorems are demonstratively certain; their truth is established by demonstrating that they are deductive consequences of the axioms. Any appeal to the measurement of figures drawn on paper or in sand is wholly irrelevant. Hume declared that "though there never were a circle or triangle in nature, the truths demonstrated by Euclid would for ever retain their certainty and evidence."[11]

The truth or falsity of statements about matters of fact, on the other hand, must be established by an appeal to empirical evidence. One cannot establish the truth of a statement that something has happened, or will happen, simply by thinking about the meaning of words.

Hume thus effected a demarcation of the necessary statements of mathematics from the contingent statements of empirical science, thereby sharpening Newton's distinction between a formal deductive system and its application to experience. Albert Einstein later rephrased Hume's insight as follows: "as far as the laws of mathematics refer to reality, they are not certain; and as far as they are certain, they do not refer to reality."[12] Hume's demarcation placed a roadblock in the path of any naïve Pythagoreanism which seeks to read into nature a necessary mathematical structure.

The Principle of Empiricism

Hume maintained that Descartes was wrong to hold that we possess innate ideas of mind, God, body, and world. According to Hume sense impressions are the sole source of knowledge of matters of fact.* He thus echoed Aristotle's dictum that there is nothing in the intellect which was not first in the senses. Hume's version was that "all our ideas are nothing but copies of our impressions, or, in

---

* Hume included among "sense impressions" desires, volitions, and feelings, as well as visual, auditory, tactile, and olfactory data.

other words, that it is impossible for us to think of any thing, which we have not antecedently felt, either by our external or internal senses."[13]

Hume's thesis is both a psychological hypothesis about the genesis of empiricial knowledge and a logical stipulation of the range of empirically significant concepts. Hume restricted empirically significant concepts to those which can be "derived from" impressions.[14] Thus stated, Hume's criterion is quite vague. Elsewhere in the *Enquiry*, he suggested that the role of the mind in generating knowledge is restricted to the compounding, transposing, augmenting, or diminishing, of the ideas "copied from" impressions.[15] Presumably, any concept is excluded which is neither a "copy" of an impression nor the result of a process of compounding, transposing, augmenting, or diminishing. Concepts excluded by Hume himself include "a vacuum",[16] "substance",[17] "perduring selfhood",[18] and "necessary connectedness of events".[19]

Hume's analysis has been interpreted as reinforcing Baconian inductivism, a tradition that perhaps owes as much to Hume's epistemological investigations as to the counsel of Francis Bacon himself. Thus interpreted, Hume has been held to claim that science begins with sense impressions and can encompass only those concepts which are "constructed" somehow out of sense data. Such a view is consistent with the Method of Analysis, but not with Newton's axiomatic method.

But although this reading of Hume has been influential it fails to do justice to the complexity of Hume's position. For Hume acknowledged that the formulation of comprehensive theories, such as Newton's mechanics, is achieved by a creative insight not reducible to a "compounding, transposing, augmenting, or diminishing" of ideas "copied from" impressions. What he did deny, however, is that any such theories could achieve the status of necessary truth.

Analysis of Causation

Bacon and Locke had discussed the question of a necessary knowledge of nature from a scholastic standpoint. Both had recommended the study of the coexistence of properties. Hume shifted the search for necessary empirical knowledge to sequences of events. He asked whether a necessary knowledge of such sequences was possible, and decided that it was not. Hume held that to establish a necessary knowledge of a sequence of events one would have to prove that the sequence could not have been otherwise. But Hume pointed out that it was not a self-contradiction to affirm that although every *A* has been followed by a *B*, the next *A* will not be followed by a *B*.

Hume undertook to examine our idea of a "causal relation". He noted that if we mean by a 'causal relation' both 'constant conjunction' and 'necessary connection', then we can achieve no causal knowledge at all. This is because we have no impression of any force or power by means of which an *A* is constrained to produce a *B*. The most that we can establish is that events of one type invariably have been followed by events of a second type. Hume concluded that the only "causal" knowledge that we can hope to achieve is a knowledge of the *de facto* association of two classes of events.

Hume conceded that we do feel that there is something necessary about many sequences. According to Hume, this feeling is an impression of the "internal sense", an impression derived from custom. He declared that "after a repetition of similar instances, the mind is carried by habit, upon the appearance of one event, to expect its usual attendant, and to believe that it will exist."[20] Of course, the fact that the mind comes to anticipate a *B* upon the appearance of an *A* is no proof that there is a necessary connection between *A* and *B*.

Consistent with this analysis, Hume stipulated definitions of 'causal relation' both from an objective and from a subjective standpoint. Objectively considered, a causal relation is a constant conjunction of the members of two classes of events; subjectively considered, a causal relation is a sequence such that, upon appearance of an event of the first class, the mind is led to anticipate an event of the second class.

These two definitions appear both in the *Treatise* and in the *Enquiry*.[21] However, in the *Enquiry*, Hume inserted after the first definition the following qualification: "or in other words where, if the first object had not been, the second never had existed."[22] Replacing the term 'object' by 'event', which is consistent with Hume's own usage, it is evident that this new definition is not equivalent to the first definition. For instance, in the case of two similar pendulum clocks arranged to be 90 degrees out of phase, the ticks of the two clocks are constantly conjoined, but this does not imply that if the pendulum of clock 1 were arrested, then clock 2 would cease to tick.

Hume's inclusion of this qualification in the *Enquiry* may indicate that he was not quite satisfied to equate causal relation and *de facto* regularity. Another likely indication of his uneasiness is the fact that he included in the *Treatise*, tersely and without comment, a list of eight "Rules by which to judge of Causes and Effects".[23] Among these rules are versions of the Methods of Agreement, Difference, and Concomitant Variations, later made famous by Mill.

The Method of Difference, in particular, enables the investigator to judge causal connection upon observation of just two instances. It would seem, in this case, that Hume contradicted his "official position" that we term a relation "causal" only upon experience of a constant conjunction of two types of events. Hume denied this. He maintained that although belief that a succession of events is a causal sequence may arise even after a single observation of the sequence, the belief nevertheless is a product of custom. This is because the judgement of causal connection in such cases depends implicitly on the generalization that like objects in like circumstances produce like effects. But this generalization itself expresses our expectation based on extensive experience of constantly conjoined events. Hence our belief in a causal connection invariably is a matter of habitual expectation.

Having thus accounted for the *origin* of our belief in causal connection, Hume was quick to point out that no appeal to the regularity of past experience can guarantee fulfilment of our expectations about the future. He stated that "it is impossible, therefore, that any arguments from experience can prove this resemblance of the past to the future; since all these arguments are founded on the supposition of that resemblance."[24] Hence it is not possible to achieve a demonstrative knowledge of causes from premises which state matters of fact.

Hume thus completed a sweeping attack on the possibility of a necessary knowledge of nature. Such knowledge would have to be either immediate or demonstrative. Hume had shown that no immediate knowledge of causes is possible, for we have no impression of necessary connection. He also had shown that it is not possible to achieve a demonstrative knowledge of causes, either from premises which state *a priori* true relations of ideas, or from premises which state matters of fact. There seemed to be no further possibility. No scientific interpretation can achieve the certainty of a statement such as 'the whole is greater than each of its parts.' Probability is the only defensible claim that can be made for scientific laws and theories.

Although Hume's scepticism was apprehended as a threat to science by those who were not satisfied with "merely probable" knowledge, Hume himself was quite ready to rely on the testimony of past experience. On the practical level, Hume was not a sceptic. He declared that

custom, then, is the great guide of human life. It is that principle alone which renders our experience useful to us.... Without the influence of custom, we should be entirely ignorant of every matter of fact beyond what is immediately present to the memory and senses.[25]

Kant on Regulative Principles in Science

Response to Hume

Immanuel Kant professed to be greatly disturbed by Hume's analysis of causation. Kant conceded that if the form and content of scientific laws wholly derive from sense experience, as Hume had urged, then there is no escape from Hume's conclusion. However, Kant was unwilling to grant Hume's premiss. Against Hume, he argued that although all empirical knowledge "arises from" sense impressions, it is not the case that all such knowledge is "given in" these impressions. Kant distinguished between the matter and the form of cognitive experience. He held that sense impressions provide the raw material of empirical knowledge, but that the knowing subject itself is responsible for the structural-relational organization of this raw material.

Kant believed that Hume had oversimplified the knowing process by reducing the operations of the mind to a mere "compounding, transposing, augmenting, and diminishing" of ideas "copied from" impressions. Kant's own theory of knowledge was more complex. He specified three stages in the cognitive organization of experience. First, unstructured "sensations" are ordered with respect to Space and Time (the "Forms of the Sensibility"). Second, the "perceptions" thus ordered are related by means of such concepts as Unity, Substantiality, Causality, and Contingency (four of the twelve

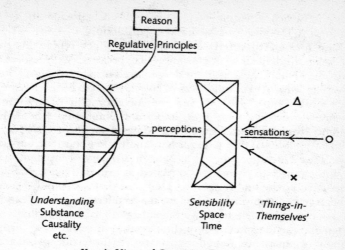

Kant's View of Cognitive Experience

"Categories of the Understanding"). Third, the "judgements of experience" thus formed are organized into a single system of knowledge through application of "Regulative Principles of Reason".

According to Kant, Hume's inadequate theory of knowledge was associated with an equally inadequate theory of science. Kant believed that Hume was preoccupied with inductive generalization. Kant held that this emphasis draws attention from the most important feature of science—the attempt to achieve a systematic organization of knowledge. Kant was profoundly impressed by the scope and power of Euclidean geometry and Newtonian mechanics and he attributed this scope and power to the deductive structure of these disciplines.

Kant regarded the systematic organization of experience as a goal to be sought by the knowing subject. He believed that progress towards the desired systematization is achieved through the application of regulative principles. In Kant's theory of knowledge, the faculty of Reason prescribes to the Understanding certain rules for the ordering of empirical judgements. Kant was quite clear that the regulative principles of Reason cannot be used to justify any particular system of empirical judgements. Rather, they prescribe ways in which scientific theories may be constructed so as to conform to the ideal of systematic organization.

Kant formulated criteria of acceptability which reflect this emphasis on the systematic organization of experience. With respect to individual empirical laws, Kant downplayed instance-confirmation, in which deductive consequences of laws are seen to be in agreement with observations. He believed the incorporation of laws into deductive systems to be more important. Kant would hold, for instance, that although Kepler's laws do gain support from data on planetary motions, they gain further, and more important, support from their "incorporation" into Newton's theory of mechanics.

With respect to theories, Kant cited as criteria of acceptability predictive power and testability. He noted that successful theories bind together empirical laws by means of reference to new entities or relations. Implicit in this systematization is the possibility of extending the interpretation of these entities or relations to further regions of experience. Kant drew attention to the fertility of scientific theories. He suggested that those theories are most acceptable which extend our knowledge of relations among phenomena.

The Analogies of Experience and the Science of Mechanics

In the *Critique of Pure Reason*, Kant singled out three "analogies of experience" which are associated with the Categories of Substance, Causality, and Interaction. He maintained that these analogies stipulate necessary conditions of the very possibility of objective empirical knowledge. The first analogy—the principle of the permanence of substance—specifies that substance is conserved throughout all changes. The second analogy—the principle of causality—specifies that for every event, there is some set of antecedent circumstances from which the event follows according to a rule. And the third analogy—the principle of community—specifies that substances perceived as coexistent in space are in interaction with one another.

In the *Metaphysical Foundations of Natural Science*, Kant sought to explain how these analogies apply to the science of mechanics. According to Kant, the subject-matter of mechanics is matter in motion, in so far as this matter possesses attractive and repulsive forces. He held that, as applied to mechanics, the analogies of experience are transformed into the principles of conservation of matter, inertial motion, and equality of action and reaction, viz.:

Category	Analogy of Experience	Principle of Mechanics
Substance	Conservation of Substance	Conservation of Matter
Causality	Principle of Causality (Every event has an antecedent from which it follows in accordance with a rule)	Principle of Inertia (All changes of motion of bodies result from extrinsic forces)
Interaction	Community of Interaction (All things that exist simultaneously are reciprocally related)	Equality of Action and Reaction

Kant maintained that the three principles of mechanics are regulative principles that should guide the search for specific empirical laws. These principles stipulate that to explain an event one must find a set of prior circumstances from which events of the same type follow according to a rule, in such a way that matter is conserved, changes in the motion of a body are attributed to forces extrinsic to the body itself, and action is balanced by reaction. Kant insisted that objective empirical knowledge can be achieved only if individual laws are formulated so as to conform to these principles.

Systematic Organization of Empirical Laws

Kant held that there are further regulative principles that apply to the organization of individual laws into a systematic interpretation of nature. In the *Critique of Judgement* (1790), he declared that

the reflective judgment, which is obliged to ascend from the particular in nature to the universal, requires on that account a principle that it cannot borrow from experience, because its function is to establish the unity of all empirical principles under higher ones, and hence to establish the possibility of their systematic subordination. Such a transcendental principle, then, the reflective judgment can only give as a law from and to itself.[26]

According to Kant, the general regulative principle which the reflective judgement prescribes to itself is the Purposiveness of Nature.

Kant insisted that although we cannot prove that nature is purposively organized, we must systematize our empirical knowledge by viewing nature *as if* it were so organized. Kant believed that systematization of empirical knowledge is possible only if we act on the presupposition that an "understanding" other than our own has furnished us with particular empirical laws so arranged as to make possible for us a unified experience.

In itself, the Principle of the Purposiveness of Nature appears to tell us only that if we seek to construct a systematic subordination of empirical laws, we must act on the assumption that such an achievement is possible. Presumably we may exclude inconsistent sets of laws as incompatible with a purposive organization of nature. But this provides but a small clue as to what types of system would satisfy the Principle of Purposiveness.

Kant further specified the meaning of the Principle of Purposiveness by formulating a list of presuppositions which he believed to be suggested by that principle:

1. that nature takes the shortest way (*lex parsimoniae*);*
2. that nature "makes no leaps either in the course of its changes or in the juxtaposition of specifically different forms (*lex continui in natura*)";
3. that there exists in nature only a small number of types of causal interaction;
4. that there exists in nature a subordination of species and genera comprehensible by us; and
5. that it is possible to incorporate species under progressively higher genera.[27]

These presuppositions become regulative principles when the investigator interrogates nature on the assumption that the presuppositions are fulfilled. Kant held that these regulative principles specify how we *ought* to judge in order to achieve a systematic knowledge of nature.[28]

In the *Critique of Pure Reason*, Kant suggested three additional

* Kant was much impressed by Maupertuis's principle of least action, a principle from which—upon suitable interpretation of 'action'—laws governing static equilibrium, collisions, and refraction could be derived. The principle of least action, like Leibniz's principle of least effort, appeared to provide a reason why these individual laws are obeyed. Maupertuis interpreted the principle as evidence of the purposive activity of the Creator. Kant, however, attributed to the principle only the status of a regulative principle.

regulative principles to guide research in the taxonomic disciplines: a Principle of Homogeneousness, which stipulates that specific differences be disregarded so that species may be grouped into genera; a Principle of Specification, which stipulates that specific differences be emphasized so that species may be divided into subspecies; and a Principle of the Continuity of Forms, which stipulates that there be a continuous, gradual transition from species to species. Kant maintained that the Principle of Homogeneousness is a check against finding an extravagant variety of species and genera, that the Principle of Specification is a check against hasty generalization, and that the Principle of the Continuity of Forms unites the first two principles by requiring that a balance be struck between them.[29]

In addition to prescribing these various regulative principles, Kant defended the use of idealizations in scientific theories. He recognized that in many cases the systematic organization of empirical laws is facilitated by the introduction of conceptual simplification. Hence he did not wish to limit the raw material of scientific theories to concepts "derived from nature". Kant cited the concepts "pure earth", "pure water", and "pure air" as examples of idealizations not inferred from phenomena, and suggested that the use of such concepts facilitates the systematic explanation of chemical phenomena.[30] Kant's examples are less forceful than Galileo's expressly formulated idealizations "ideal pendulum" and "free fall in a vacuum", but Kant must be credited with the insight that a naïve empiricism fails to provide a sufficiently rich conceptual basis for science.

Teleological Explanations

The Principle of Purposiveness enjoins us to investigate nature *as if* the laws we discover were part of a system of laws arranged by an "understanding" other than our own. If we proceed on this basis, we are bound to inquire about the place of particular laws in the system of nature as a whole. This is particularly true in the biological sciences. We cannot help but ask questions about the purposes served by observed patterns of structure, function, and behaviour. Answers to such questions often are teleological explanations, characterized by use of the phrase 'in order that' or its equivalent.

Kant believed that teleological explanations were of value in science for two reasons. In the first place, teleological explanations are of heuristic value in the search for causal laws. Kant maintained that asking questions about "ends" may suggest new hypotheses about "means", thereby extending our knowledge of the mechanical interaction of systems and their parts.[31] In the second place, teleological interpretations contribute to the ideal of the systematic organization of empirical knowledge by supplementing the available causal interpretations. Kant believed that causal interpretations should be extended as far as possible, but he was pessimistic about the possibility of an extensive causal interpretation of life processes.

Kant's pessimism was based on his conception of the nature of living organisms. According to Kant, living organisms exhibit a reciprocal dependence of part and whole; not only is the whole what it is in virtue of an organization of parts, but also a part is what it is in virtue of its relation to the whole. Each part of a living organism is related to the whole both as cause and as effect. An organism is both an organized whole and a self-organizing whole. Kant believed that this reciprocal dependence of part and whole cannot be explained fully by causal laws. Causal laws establish only that particular states of an organism follow from other states according to a rule.

There are limitations, therefore, on a causal interpretation of nature. Kant set forth the limitations, but he did not counsel a return to an "easy teleology", in which the structures and functions of organisms are dismissed by reference to "final causes". For Kant, the proper explanation of natural phenomena is in terms of laws which state patterns according to which events occur. The concept of causality is constitutive of objective empirical knowledge; the concept of purpose is not. Kant maintained that purposiveness can be only a regulative principle by means of which Reason selects as its goal the systematic organization of empirical laws. By relocating teleology at the level of the regulative activity of Reason, Kant achieved the integration of teleological and mechanistic emphases that Leibniz had sought.

Notes

[1] John Locke, *An Essay Concerning Human Understanding*, IV, iii. 25.
[2] Ibid. II. viii. 23.

[3] Ibid. IV. vi. 14.

[4] John Yolton, *Locke and the Compass of Human Understanding* (Cambridge: Cambridge University Press, 1970), 58.

[5] Locke, *Essay*, IV. xii. 10.

[6] G. W. Leibniz, 'On a General Principle Useful in Explaining the Laws of Nature through a Consideration of the Divine Wisdom; To Serve as a Reply to the Response of the Rev. Father Malebranche', in L. Loemker, ed., *Leibniz: Philosophical Papers and Letters* (Dordrecht: D. Reidel Publishing Co., 1969), 351–3.

[7] Leibniz, 'Tentamen Anagogicum: An Anagogical Essay in the Investigation of Causes', *Leibniz: Philosophical Papers and Letters*, 477–84.

[8] Gerd Buchdahl, *Metaphysics and the Philosophy of Science* (Oxford: Blackwell, 1969), 416–17.

[9] Leibniz, 'Seventh Letter to de Volder (November 10, 1703)'; 'Eighth Letter to de Volder (January 21, 1704)'; in *Leibniz: Philosophical Papers and Letters*, 533. See also George Gale, 'The Physical Theory of Leibniz', *Studia Leibnitiana II*, 2 (1970), 114–27.

[10] See Leibniz, 'Sixth Letter to de Volder (June 20, 1703)', in *Leibniz: Philosophical Papers and Letters*, 530.

[11] David Hume, *An Enquiry Concerning Human Understanding* (Chicago: Open Court Publishing Co., 1927), 23.

[12] Albert Einstein, 'Geometry and Experience' in *Sidelights on Relativity* (New York: E. P. Dutton Co., 1923), 28.

[13] Hume, *Enquiry Concerning Human Understanding*, 63.

[14] Ibid. 19.

[15] Ibid. 16.

[16] Hume, *A Treatise of Human Nature*, 53–65.

[17] Ibid. 15–16.

[18] Ibid. 251–62.

[19] Ibid. 155–72.

[20] Hume, *Enquiry Concerning Human Understanding*, 77.

[21] Hume, *Treatise of Human Nature*, 172; *Enquiry Concerning Human Understanding*, 79.

[22] Hume, *Enquiry Concerning Human Understanding*, 79.

[23] Hume, *Treatise of Human Nature*, 173–5.

[24] Hume, *Enquiry Concerning Human Understanding*, 37.

[25] Ibid. 45.

[26] Immanuel Kant, *Critique of Judgement*, trans. J. H. Bernard (London: Macmillan, 1892), 17.

[27] Ibid. 20–4.

[28] Ibid. 21.

[29] Kant, *Critique of Pure Reason*, trans. F. Max Müller (New York: Macmillan, 1934), 530.

[30] Ibid. 519.

[31] Kant, *Critique of Judgement*, 327.

II. THEORIES OF SCIENTIFIC PROCEDURE

JOHN HERSCHEL (1792–1871) was the son of the great astronomer William Herschel. The achievements of the elder Herschel included the discovery of Uranus and the compilation of valuable data on double stars and nebulae.

John Herschel studied at Cambridge, and thereafter devoted his life to the pursuit of science. His scientific achievements included studies of double refraction in crystals, experiments in photography and photochemistry, a method of computing binary-star orbits, and numerous astronomical observations. Herschel spent the period 1834–8 at the Cape of Good Hope, where he successfully extended his father's survey of double stars and nebulae to the Southern skies.

Herschel published *A Preliminary Discourse on the Study of Natural Philosophy* in 1830. His analysis of the role of hypothesis, theory, and experiment in science was acknowledged to be influential by Whewell, Mill, and Darwin, among others.

WILLIAM WHEWELL (1794–1866) graduated from Trinity College, Cambridge, where he was appointed Professor of Minerology (1828), Professor of Moral Philosophy (1838), and Vice-Chancellor (1842). He was instrumental in introducing into England the Continental version of the calculus, and was largely responsible for broadening the course of study at Cambridge.

Whewell performed extensive researches on the tides, and was recognized—by Lyell and Faraday, among others—as an authority on scientific nomenclature. He completed his extensive *History of the Inductive Sciences* in 1837, and based his *Philosophy of the Inductive Sciences* (1840) on the results of this historical analysis.

ÉMILE MEYERSON (1859–1933) was born in Lublin, Russian Poland, studied at various European universities, and then combined research into the history and philosophy of science with the practice of chemistry in France. Meyerson viewed the history of science as a continuing search for that which is conserved throughout change. His published works include *Identity and Reality* (1907), and studies of quantum mechanics and the theory of relativity.

John Herschel's Theory of Scientific Method

John Herschel's *Preliminary Discourse on Natural Philosophy* (1830), was the most comprehensive and best-balanced work on the philosophy of science available at that time. Herschel was one of the foremost English scientists of his day, and his writing on scientific method was distinguished by careful analyses of recent achievements in physics, astronomy, chemistry, and geology.

One of Herschel's important contributions to the philosophy of science was a clear distinction between the "context of discovery" and the "context of justification". He insisted that the procedure used to formulate a theory is strictly irrelevant to the question of its acceptability. A meticulous inductive ascent and a wild guess are on the same footing if their deductive consequences are confirmed by observation.

Context of Discovery

Although he respected Francis Bacon's views on scientific inquiry, Herschel was aware that many important scientific discoveries do not fit the Baconian pattern. For this reason, he maintained that there are two distinct ways in which a scientist may proceed from observations to laws and theories. One approach is the application of specific inductive schema. The other is the formulation of hypotheses. Herschel's view of the context of discovery may be represented schematically as follows:

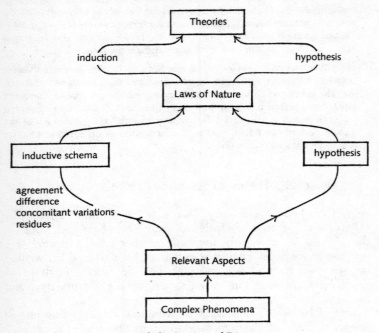

Herschel's Pattern of Discovery

According to Herschel, the first step in scientific procedure is to subdivide complex phenomena into their constituent parts or aspects, and to fix attention on those properties which are crucial for explaining the phenomena. To account for the motion of bodies, for example, one must focus on such properties as force, mass, and velocity. Herschel's principal example of the reduction of a complex phenomenon into its relevant aspects is the analysis of sound into the vibration of a source, the transmission of vibratory motion through a medium, its reception by the ear, and the production of sensation. He held that a complete understanding of sound would require a knowledge of impact phenomena which issue in vibration, a knowledge of the interaction of a moving particle and the particles which surround it, and a knowledge of the physiology of auditory sensations.[1]

Laws of Nature. Phenomena properly analysed are the raw material from which the scientist seeks to formulate "laws of nature". Herschel included among laws of nature both correlations of properties and sequences of events. Among lawful correlations of properties are Boyle's Law and the generalization that doubly refracting substances exhibit periodical colours under polarized light. Herschel spoke of such correlations as "general facts". Among lawful sequences of events are Galileo's laws of free fall and the parabolic trajectory of projectiles.

Herschel noted that laws of nature are affirmed implicitly with a stipulation that certain boundary conditions are fulfilled. For instance, the law of free fall is affirmed to hold only for motion in a vacuum, and Boyle's Law is affirmed to hold only for changes at constant temperature.

Herschel traced two distinct routes from phenomena to laws of nature. The first route to the discovery of laws is by application of specific inductive schema. Boyle's Law, for example, was discovered by studying the variation of the volume of a gas with its pressure, and generalizing from the experimental results. For example, given the data:

P	V
0.5	2.0
1.0	1.0
2.0	0.5
5.0	0.2

the investigator may conclude that $P \propto (1/V)$.

The second route to the discovery of laws is by formulation of hypotheses. Herschel emphasized that this latter route to laws of nature cannot be reduced to the application of fixed rules. He cited as an example Huygens's hypothesis that the extraordinary ray in doubly-refracting Iceland spar is propagated elliptically. Even though Huygens had no conception of the transverse wave motion of light, he was able to formulate a law which accounts for double refraction by means of this hypothesis of elliptic propagation. According to Herschel, Huygens's hypothesis cannot be represented as the conclusion of an inductive schema.[2]

Theories. The discovery of laws of nature is only the first stage in scientific interpretation. The second stage is the incorporation of these laws into theories. According to Herschel, theories arise either upon further inductive generalization, or by creation of bold hypotheses that establish an interrelation of previously unconnected laws.

Herschel combined the Baconian ideal of a hierarchy of scientific generalizations with a perceptive emphasis on the role of the creative imagination in the construction of the hierarchy. One imaginative theory which impressed him was Ampère's theory of electromagnetism. Ampère explained the mutual attraction or repulsion of magnets by positing the existence of circulating electric currents within the magnets. Ampère did not arrive at this theory upon application of inductive schema to the laws of electricity and magnetism. However, the theory does have testable consequences, and Herschel insisted that its acceptability is determined, not by the method of its formulation, but by the experimental confirmation of these consequences.[3]

Context of Justification

Herschel emphasized that agreement with observations is the most important criterion of acceptability for scientific laws and theories. Moreover, he insisted that some confirming instances are of greater significance than others.

One important type of confirming instance is the extension of a law to extreme cases. Herschel noted, for example, that the identical acceleration of a coin and a feather in an experimentally produced vacuum was a "severe test" of Galileo's law of falling bodies.[4]

A second important type of confirming instance is an unexpected result which indicates that a law or theory has an undesigned scope. Herschel declared that

the surest and best characteristic of a well-founded and extensive induction . . . is when verifications of it spring up, as it were, spontaneously, into notice, from quarters where they might be least expected, or even among instances of that very kind which were at first considered hostile to them.[5]

He noted, for example, that discovery of the elliptic orbits of binary star systems was unexpected confirmation of Newtonian mechanics,[6] and that the existence of a discrepancy between calculated and observed velocities of sound was an unexpected confirmation of the law of heat generation by compression of an elastic fluid.[7]

A third important type of confirming instance is the "crucial experiment". Herschel regarded crucial experiments as destruction tests which acceptable theories must survive.

He cited with admiration an experiment which had been suggested by Francis Bacon to determine whether the downward acceleration of bodies is the result of attraction of the Earth or of some mechanism internal to the bodies themselves. Bacon had suggested that the issue be decided by comparing the behaviour of a weight-driven clock and a spring-driven clock at high altitudes and in mines.[8]

In addition, Herschel credited Pascal with having designed a crucial experiment to decide whether the rise of mercury in closed tubes is the result of atmospheric pressure or of "abhorrence of a vacuum". According to Herschel, Pascal's comparison of the heights of a mercury column at the base and top of a mountain refuted the "abhorrence" hypothesis and left Torricelli's "sea of air" hypothesis alone in possession of the field.[9]

It may be objected that, whereas the proposed experiments of Bacon and Pascal may provide striking confirmation of particular hypotheses, they properly are termed "crucial" only if every possible alternative hypothesis is inconsistent with the results obtained. Failure to give due weight to this requirement led Herschel, and many other nineteenth-century scientists, to accept Foucault's determination that the velocity of light is greater in air than in water as a "crucial" experiment. Foucault's result was consistent with Huygens's wave theory, but was inconsistent with Newton's corpuscular theory. Many scientists concluded from this that light must be "really" a wave. The implicit assumption that these two theories are the only possible interpretations of optical phenomena later proved to be incorrect.

Despite the fact that too much significance has been attributed to certain experiments in the evaluation of competing theories, the general attitude which promotes a search for falsifying instances has been most important in the history of science. Herschel encouraged this attitude. He demanded that the scientist assume the role of antagonist against his own theories, and seek both direct refutations and exceptions which limit the range of application of these theories. Herschel believed that the worth of a theory is proved only by its ability to withstand such attacks.

Whewell's Conclusions about the History of the Sciences

Morphology of Scientific Progress

William Whewell, a contemporary of Herschel, sought to base his philosophy of science on a comprehensive survey of the history of science. Whewell proposed to examine the actual process of discovery in the various sciences in order to see if any patterns are displayed therein.

Whewell claimed originality for his approach, pointing out that previous writers on the philosophy of science had regarded the history of science as a mere storehouse of examples which may be cited to illustrate particular points about scientific method. Whewell proposed to invert this relationship which had made the history of science dependent on the philosophy of science.

Whewell was quite sophisticated about the methodology of historical research. He recognized that recovery of the past necessarily involves acts of synthesis on the part of the historian. Accordingly, he selected certain interpretative categories to guide his historical studies. Whewell saw scientific progress as a successful union of facts and ideas, and took the polarity of fact and idea to be the basic methodological principle for the interpretation of the history of science. Armed with this principle, he sought to show the progress of each science by tracing the discovery of its pertinent facts and the integration of these facts under appropriate ideas.

Facts and Ideas. Whewell sometimes spoke of "facts" as reports of our perceptual experience of individual objects. However, he insisted that this was just one kind of fact. Broadly considered, a fact is any piece of knowledge which is raw material for the formulation of laws and theories. From this point of view, Kepler's Laws were facts upon which Newton theorized. Whewell held that there is only a relative distinction between fact and theory. If a theory is incorporated within another theory, it becomes a fact in its own right.

Whewell termed "ideas" those rational principles which bind together facts. Ideas express the relational aspects of experience which are a necessary condition for understanding. Whewell affirmed Kant's thesis that ideas are prescribed to, and are not derived from, sensations. Whewell included among ideas both general notions such as space, time, and cause, and ideas basic to particular sciences. Examples of the latter are "elective affinity" in chemistry, "vital forces" in biology, and "natural types" in taxonomy.

Whewell conceded that there can be no such thing as a "pure fact" divorced from all ideas. Any fact about an object or process necessarily involves the ideas of space, time, or number. Consequently, even the simplest facts involve something of the nature of theory. Whewell's distinction between fact and theory is at bottom a psychological distinction. When we label something a 'fact', we usually are unaware of the way in which relational principles integrate our sense experience. For example, we take it to be a fact that a year is approximately 365 days. But this fact involves the ideas of time, number, and recurrence. We call this relation a 'fact' only because we do not attend to the associated ideas. By contrast, when we label something a 'theory', our attention is directed to the ideas applied to integrate facts. Whewell declared that 'we still have an intelligible distinction of Fact and Theory, if we consider Theory as a conscious, and Fact as an unconscious inference, from the phenomena which are presented to our senses.'[10] He believed that the concepts 'fact', 'idea', and 'theory', are of value for interpreting the history of science, even though every theory may be also a fact and every fact partakes of the nature of theory.

Pattern of Scientific Discovery. The pattern of scientific discovery which Whewell claimed to see in the history of the sciences was a three-beat progression comprising a prelude, an inductive epoch, and a sequel. The prelude consists of a collection and decomposition of facts, and a clarification of concepts. An inductive epoch arises when a particular conceptual pattern is superinduced on the facts. And its sequel is the consolidation and extension of the integration thus achieved. This pattern of discovery may be schematized as in the figure overleaf:

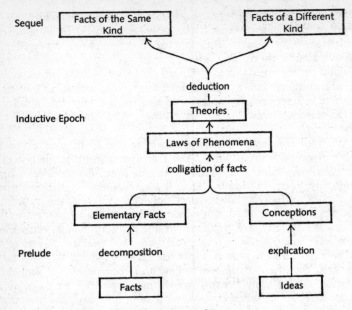

Whewell's Pattern of Discovery

Although Whewell claimed that this pattern is repeated in the history of the sciences, he was careful to point out that the stages within the pattern often overlap. Within the history of a particular science, the explication of conceptions may accompany, as well as precede, the formulation of laws, and the formulation of theories, may accompany, as well as precede, the verification of laws. Nevertheless, he claimed to have represented, by this pattern, the morphology of scientific progress.

Decomposition of Facts and Explication of Conceptions. Whewell held that the decomposition of facts and the explication of conceptions are necessary stages in theory-construction. The decomposition of facts is a reduction of complex facts to "elementary" facts which state relations among such clear and distinct ideas as space, time, number, and force. In many instances this is achieved by focusing on qualities which undergo quantitative variation, and by developing techniques for recording values of these qualities.

The notion of the explication of conceptions is more difficult to pin down. Within the history of science, discussions among scientists often result in the clarification of concepts. Whewell noted that it was through such discussions that the concepts of "force", "polarization", and "species" have been clarified, and he called for a similar clarification of the concept of "life".

One difficulty about Whewell's notion of explication is the nature of the clarification achieved. Whewell spoke of conceptions as "special modifications" of the fundamental ideas of the sciences.[11] As such, conceptions have a less extensive range of application than do the fundamental ideas themselves. Whewell included among conceptions "accelerating force" and "neutral combination of elements".[12] He held that such conceptions are explicated when their logical relations to the fundamental ideas are clearly recognized.

Whewell believed that the meaning of a fundamental idea may be expressed by a set of axioms which state basic truths about the idea. He maintained that a derivative conception is explicated only when it is related to the fundamental ideas in such a way that the "necessary cogency" of these axioms is understood. And to understand the "necessary cogency" of the axioms is "clearly and steadily" to contemplate the idea itself.[13]

The inevitable question at this point is how to recognize that a scientist has achieved a "clear and steady" apprehension of an idea. Of course, in retrospect, one can gauge the clarity of an idea by the success of the theory in which it is embedded. On this approach, one may conclude, as Whewell did, that the concept of inertia was clarified progressively in the work of Galileo, Descartes, and Newton.

Whewell maintained that, in addition to being clear, useful scientific conceptions are "appropriate" to the facts to which they are applied. He conceded that, for the most part, we can establish the appropriateness of conceptions only by pointing to confirmations of laws and theories which utilize them. Nevertheless, he thought that in certain cases the criterion of appropriateness could be used to rule out in advance misguided interpretations. For example, since the proper goal of physiology is truths respecting "vital powers", one can exclude from physiology interpretations based exclusively on mechanical principles or chemical principles.

Colligation of Facts. Whewell maintained that laws and theories are a "colligation" in which the investigator superinduces a conception upon a set of facts. He spoke of colligation as a "binding

together" of facts, and chose the formulation of Kepler's Third
Law to illustrate this process of integration. Kepler succeeded in
binding together facts about the planets' periods of revolution and
distances from the sun, by means of such conceptions as 'squares
of numbers', 'cubes of distances', and 'proportionality'.[14]

According to Whewell, Kepler's achievement was a triumph of
induction. He declared that in its proper use "Induction is a term
applied to describe the process of a true Colligation of Facts by
means of an exact and appropriate Conception".[15] Several aspects
of Whewell's discussion of induction deserve comment.

Whewell held that induction is a *process* of discovery. It is not a
schema for proving propositions. This is not to say that Whewell
was uninterested in the problem of evaluating the evidence for
inductive generalizations. But he took this to be a problem of the
"logic of induction". Induction itself is the process of generalizing
from facts in such a way that a colligation is achieved.

Whewell's examination of the history of science convinced him
that the colligation of facts is achieved through the creative insight
of scientists, and not by means of the application of specific inductive
rules. He observed that the success of induction "seems to consist
in framing several tentative hypotheses and selecting the right one.
But a supply of appropriate hypotheses cannot be constructed
by rule, nor without inventive talent."[16] According to Whewell,
induction is a process of invention and trial. He cited the example
of Kepler, who tried to fit the facts of planetary motion to numerous
ovoid orbits, before finally achieving success with the hypothesis of
elliptical orbits. In addition, Whewell listed a number of cases of
"felicitous and inexplicable strokes of inventive talent" in the history
of science.[17]

Whewell's principal thesis about induction is that the process of
scientific discovery cannot be reduced to rules. However, he did
recognize that considerations of simplicity, continuity, and symmetry
often are affirmed as regulative principles in the selection of
hypotheses. Whewell also suggested that specific inductive methods,
such as the method of least squares and the method of residues, are
of value in the formulation of mathematically quantified laws.

A corollary of Whewell's position on induction and hypothesis is
that an inductive inference is always something more than a mere
collection of facts. Whewell stated that "the Facts are not only
brought together, but seen in a new point of view. A new mental

Element is superinduced; and a peculiar constitution and discipline of mind are requisite in order to make this Induction."[18]

Tributary–River Analogy. Whewell compared the evolutionary development of a science to the confluence of tributaries to form a river.[19] He concluded from his historical studies that a science evolves through the progressive incorporation of past results in present theories. He cited Newton's theory of gravitational attraction as the paradigm of this growth by incorporation. Newton's theory subsumed Kepler's Laws, Galileo's Law of Free Fall, the motions of the tides, and diverse other facts.

Whewell was aware that successive interpretations of particular phenomena are not always consistent. Despite this, he concluded that science was a continuing progression, rather than a series of revolutions. His emphasis was on those aspects of rejected theories which facilitated subsequent theory-formation. For example, he conceded that Lavoisier's Oxygen Theory had supplanted the Phlogiston Theory, and that many facts which are explained by the Oxygen Theory are inconsistent with the Phlogiston Theory, but he contended that the Phlogiston Theory nevertheless had played a positive role in the history of chemistry, because this theory classified together the processes of combustion, acidification, and respiration.[20] On Whewell's view, a theory contributes to scientific progress if it binds together, even for the wrong reasons, facts which indeed are related.

Consilience of Inductions

Whewell claimed that the history òf science reveals a clue for a "logic of induction". This clue is the tributary–river analogy. He concluded that, because scientific progress is a successive incorporation of laws into theories, an acceptable set of generalizations within a particular science ought to exhibit a certain structural pattern. This pattern is an "Inductive Table" which has the form of the tributary–river relation. The Inductive Table is an inverted pyramid, with specific facts at the top and generalizations of the broadest scope at the bottom. Transition from the top to the bottom of the table reflects progressive inductive generalization, in which observations and descriptive generalizations are subsumed under theories of increasing scope.

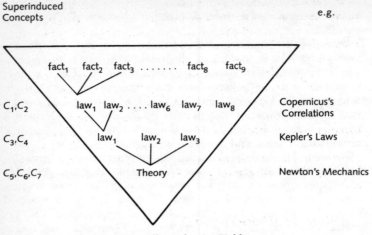

Whewell's Inductive Table

Whewell maintained that the Inductive Table specifies the form of a valid set of inductive inferences, in much the same way as the syllogism specifies the form of valid deductive inferences. However, he was careful not to overextend the analogy. He noted that whereas the syllogistic forms are schemata which are converted into valid deductive arguments upon the insertion of class names, the form of the Inductive Table is incomplete as a schema for construction of valid inductive inferences. This is because generalizations at one level are not simply conjoined to form higher generalizations. Rather, the more inclusive generalization incorporates lower-level generalizations only upon superinduction of a concept, or set of concepts. It is by means of conceptual integration, and not mere summation or enumeration, that lower-level generalizations are seen to be connected. For this reason, Whewell insisted that a *complete* Inductive Table must make reference to the specific concepts superinduced at each level of generality. For example, a table for the inductive generalization from Kepler's Laws to Newton's Laws both would display the form of an inverted pyramid, and would stipulate that the incorporation is accomplished by means of such superinduced concepts as force, inertial motion, and Absolute Space and Time.

Whewell contended that the incorporation of two or more generalizations into a more inclusive theory is itself a criterion of acceptability for scientific theories. He spoke of this incorporation as a "consilience of inductions", and declared that "No example can be pointed out, in the whole history of science, so far as I am aware, in which this Consilience of Inductions has given testimony in favour of an hypothesis afterwards discovered to be false."[21] Whether or not a consilience of inductions is achieved in a particular case depends on the adequacy of theoretical concepts to bind together two or more laws. The kinetic theory of gases is a good example of a successful consilience of inductions. The concept of Newtonian elastic collisions among molecules of a gas suffices to bind together in one theory the empirical laws of Boyle, Charles, and Graham.

Historicization of Necessary Truth

It has been indicated that Whewell interpreted the history of the sciences in terms of a Kantian distinction between the form and the content of knowledge. Scientific knowledge, for Whewell, is a binding together of facts by means of ideas. But since Whewell held that these ideas express necessary truths, it might seem that at least some scientific knowledge may achieve the status of necessary truth.

In an early work, Whewell maintained that the axioms of geometry and the fundamental laws of nature differ with regard to cognitive status. Geometrical axioms are necessary truths, the laws of the natural sciences are not.[22] Subsequently, however, he changed his mind, and insisted that some laws of the natural sciences rightly come to be recognized as necessary truths.

Whewell conceded the paradoxical nature of this claim. He agreed with Hume that no amount of empirical evidence can prove that a relationship could not be other than it is. And yet he believed that certain scientific laws have achieved necessary status.

Whewell's attempt to resolve the paradox hinged on a distinction between the *form* and the *matter* of the fundamental laws of nature. He held that Newton's laws of motion, for instance, exemplify the *form* of the Idea of Causation. But since the Idea of Causation is a necessary condition of the very possibility of objective empirical knowledge, Newton's laws must share this necessity.

According to Whewell, the meaning of the Idea of Causation may be unpacked in three axioms: (1) nothing takes place without a cause; (2) effects are proportional to their causes; and (3) reaction is equal and opposite to action. It remains for experience, however, to specify the *content* of these axioms. Experience teaches that brute matter possesses no intrinsic internal cause of acceleration, that forces are compounded in certain ways, and that certain definitions of 'action' and 'reaction' are appropriate. Newton's laws of motion express these findings. Whewell held that Newton's laws provide the proper empirical interpretation of the axioms of causation, thereby achieving the status of necessary truths.[23]

Whewell maintained that the necessary status of the fundamental laws of nature derives from their relation to those Ideas which are *a priori* necessary conditions of objective empirical knowledge. He did not specify the nature of this relation other than to appeal to the notion that such laws "exemplify" the form of the Ideas. However, he did hold that this "exemplification" takes place gradually in the historical development of the sciences. It is a matter of a progressive clarification of the relation of the most general inductive laws to the basic Ideas of the sciences. Whewell was quite certain that Newton's work established the necessary status of the general laws of mechanics. He was less certain about the other general laws of the sciences.

Meyerson on the Search for Conservation Laws

Émile Meyerson, writing in 1908, gave Whewell full credit for being the first to explain correctly the *a priori* necessity that distinguishes the fundamental laws of motion from mere empirical generalizations. Meyerson sought to extend Whewell's analysis by subdividing scientific laws into "empirical laws" and "causal laws".

According to Meyerson, an empirical law specifies how a system is altered when appropriate conditions are modified. Laws of this type enable us to predict the outcome of natural processes and to manipulate these processes to serve our ends. A causal law, by contrast, is an application of the Law of Identity to the existence of objects in time. It stipulates that there is something that remains the same throughout change. In the case of a chemical reaction, for instance, the atoms involved remain the same throughout the process of rearrangement.

Meyerson believed that whereas knowledge of empirical laws satisfies our demand for prevision, only knowledge of causal laws satisfies our desire for understanding. This is so in virtue of the dual aspect of causal laws. Because a causal law states an identity, it implies a necessary truth—"that which is, is, and cannot not be", as Aristotle said. But a causal law also has empirical content, since it states a claim about the existence of objects in time. It would seem that a causal law, according to Meyerson, implies both a necessary truth—the Law of Identity, and a contingent statement that a specific "substance" remains identical throughout changes of a given type. Meyerson conceded that the contingent statement may turn out to be false. This has happened, for example, in the case of the conservation of mass and the conservation of parity. Meyerson held that, in such cases, although the application of the Law of Identity to the existence of objects in time proves to be incorrect, the Law of Identity itself is unaffected.

But the Law of Identity itself is a tautology. It is not possible to deduce from it a single statement about the world. Meyerson recognized this. Nevertheless, he believed that the Law of Identity is a "significant" tautology. It is significant because the correct application of this law to the existence of objects in time is a necessary condition of an understanding of nature. The attempt to impress the Law of Identity upon nature is an important directive principle for scientific inquiry.[24]

The search for that which remains the same throughout change has been most successful in atomic theory and the conservation laws of mechanics. But, as Meyerson pointed out, the demand for identity which we impose upon nature is met with resistance at certain points. An example is Carnot's Principle, the Second Law of Thermodynamics. Carnot's Principle specifies that naturally occurring processes in an isolated system increase the entropy of the system. Entropy is a measure of degree of organization. An increase in entropy represents a decrease of organization within the system. But since there is a unidirectional increase of entropy in naturally occurring processes in isolated systems, it is not possible to regard entropy as a "substance" conserved throughout these processes. The Second Law of Thermodynamics is a relation of great scope and importance. It is a relation which is "non-causal" in Meyerson's sense. Meyerson declared that 'Carnot's principle is the expression of the resistance which nature opposes to the constraint which our understanding, through the principle of causality, attempts to exercise over it.'[25]

Notes

[1] John F. W. Herschel, *A Preliminary Discourse on the Study of Natural Philosophy* (London: Longman etc., 1830), 88–90.

[2] J. Herschel, *Familiar Lectures on Scientific Subjects* (New York: George Routledge and Sons, 1871), 362.

[3] J. Herschel, *Preliminary Discourse*, 202–3.

[4] Ibid. 168.

[5] Ibid. 170.

[6] Ibid. 280.

[7] Ibid. 171–2.

[8] Ibid. 186–7.

[9] Ibid. 229–30.

[10] William Whewell, *Philosophy of the Inductive Sciences* (London: John W. Parker, 1847), i. 42.

[11] Whewell, *Novum Organon Renovatum* (London: John W. Parker & Son, 1858), 30.

[12] Ibid. 31.

[13] Ibid. 41.

[14] Ibid. 59–60.

[15] Ibid. 70.

[16] Ibid. 59.

[17] Ibid. 64.

[18] Ibid. 71.

[19] Whewell, *History of the Inductive Sciences* (New York: D. Appleton, 1859), i. 47.

[20] Ibid. ii. 267–9.

[21] Whewell, *Novum Organon Renovatum*, 90.

[22] Whewell, *Astronomy and General Physics Considered with Reference to Natural Theology* (Philadelphia: Carey, Lea and Blanchard, 1833), 164–8.

[23] Whewell, *Philosophy of the Inductive Sciences*, i. 245–54.

[24] Émile Meyerson, *Identity and Reality*, trans. K. Loewenberg (New York: Dover Publications, 1962), 402.

[25] Ibid. 286.

III. STRUCTURE OF SCIENTIFIC THEORIES

PIERRE DUHEM (1861–1916) was Professor of Physics at the University of Bordeaux (1893–1916). He made original contributions to thermodynamics, fluid mechanics, and the history and philosophy of science. His research on medieval physics established that the "scientific revolution" of the sixteenth and seventeenth centuries had important roots in the medieval work of Buridan, Orèsme, and others. This work was a valuable corrective to that myopic view of the history of science which viewed the medieval period as a period of sterile disputation. In *The Aim and Structure of Physical Theory* (1906), Duhem maintained that scientific theories are correlative devices which group together experimental laws.

NORMAN R. CAMPBELL (1880–1949) was a Cambridge-educated physicist who worked for several years under J. J. Thomson at the Cavendish Laboratory, before joining the General Electric Company as a research physicist. His principal work on the philosophy of science is the posthumously published *Foundations of Science* (1957), an augmented version of *Physics: The Elements* (1919). Campbell's study is distinguished by careful analyses of the theory of measurement and the structure of scientific theories.

MARY B. HESSE (1924–   ) is University Reader in Philosophy of Science at Cambridge University. She studied mathematics, physics, and the history and philosophy of science at London University, and has taught at the universities of London, Leeds, and, as Visiting Professor, at Yale, Minnesota, and Chicago.

Dr Hesse is engaged at present in developing a unified view of the structure of physical science, based on inductive inference, with particular reference to historical cases of the use of models and analogies.

R. HARRÉ (1927–) is University Lecturer in Philosophy of Science at Oxford University. He has studied mathematics and physics at the University of Auckland, and philosophy at Oxford University. Prior to his appointment at Oxford, he taught in Pakistan and at Birmingham and Leicester.

A vigorous critic of deductivist and positivist philosophies of science, Harré currently is engaged in a programme to redirect the methodological orientation of the social sciences.

Pure Geometry and Physical Geometry

An adequate understanding of the process of theory-construction presupposes recognition of the distinction between an axiom system and its application to experience. The construction of non-Euclidean geometries in the nineteenth century called attention to this distinction. Lobachevsky, Bolyai, and Riemann invented axiom systems which differ in important respects from the Euclidean system.

In the Euclidean system, it is assumed that exactly one parallel line can be drawn through a point not on a given straight line. Different assumptions were made in the non-Euclidean systems. Lobachevsky and Bolyai replaced the Euclidean assumption by the axiom that through a given point there are two lines parallel to a given straight line. From this axiom, and the other axioms and definitions of his system, Lobachevsky deduced the theorem that the sum of the interior angles of a triangle is always less than 180 degrees, and decreases as the areas of triangles increase. Riemann replaced the Euclidean assumption by the axiom that through a point there are no lines parallel to a given straight line. A theorem of Riemann's geometry is that the sum of the interior angles of a triangle is always greater than 180 degrees, and increases as the areas of triangles increase.

As formal deductive systems, there are no grounds for judging one of these alternatives to be superior to the others. They are consistent relative to one another. It can be shown that if Euclidean geometry is internally consistent, then the alternative non-Euclidean geometries are consistent as well.

Recognition of this fact led many thinkers to contrast the *a priori* status of the axioms and theorems of "pure geometry" with the empirically significant assertions of "physical geometry". Helmholtz, for instance, emphasized that the various systems of geometry are, in themselves, devoid of empirical content. It is only when they are conjoined with certain principles of mechanics that empirically significant propositions result. According to Helmholtz, it is necessary to specify how such terms as 'point', 'line', and 'angle' are to be measured before geometrical theorems can be applied to experience.[1]

Duhem on the Binding Together of Laws

Pierre Duhem shared Whewell's interest in the history of science and, like Whewell, sought to formulate a philosophy of science consistent with the historical record. Whewell had drawn an image of scientific progress as a confluence of tributaries to form rivers. Duhem agreed that successful theories do colligate, or bind together, experimental laws. He spoke of theories as "representing" a group of laws, and contrasted this "representative" function with an "explanatory" function that most theories are presumed to have. Theories often are held to explain phenomena by describing "the reality underlying the phenomena". Duhem criticized this view, insisting that it is the representative function alone that is of scientific value.[2]

Duhem's position that scientific theories "represent", but do not "explain", experimental laws was based on his view of the structure of theories. According to Duhem, a scientific theory consists of an axiom system and "rules of correspondence",* which correlate some of the terms of the axiom system with experimentally determined magnitudes. There may be, in addition, a picture, or model, associated with the interpreted axiom system. But this model is not part of the logical structure of the theory. The axiom system and rules of correspondence suffice for the deduction of those experimental laws which are "represented" by the theory. Consequently, the model associated with the theory plays no part in the task of predicting the results of experiments.

In the case of the kinetic theory of gases, for example, the

---

* Duhem himself did not use the phrase 'rules of correspondence' to stand for statements which link the axiom system with experimentally determined magnitudes.

axioms state relations among terms such as 'molecule', 'velocity', and 'mass'. The axiom system is linked to experience via the concept of the root-mean-square velocity of all the molecules.* Rules of correspondence correlate this root-mean-square velocity with the pressure and temperature of the gas. Duhem insisted that the kinetic theory is valuable because it binds together previously unrelated experimental laws about the macroscopic behaviour of gases. For instance, the laws attributed to Boyle, Charles, and Graham are deductive consequences of the assumptions of the theory. This is the "representative" function of the theory. He denied, however, that the model—which depicts elastic collisions between point-masses—has any explanatory function. Duhem was highly critical of Lord Kelvin's position that to "understand" a process is to visualize an underlying mechanism. According to Duhem, the model associated with a theory may have heuristic value in the search for additional experimental laws, but the model itself is not a premiss in the explanations which are given by the theory.

Duhem emphasized that a theory does not "represent" a group of laws merely by stating a conjunction of these laws. The relationship is more complex, and it allows great range to the imagination of the theorist. Of course, an acceptable theory must imply experimentally testable laws, but the fundamental assumptions of the theory may include statements about magnitudes in no way correlated with processes of measurement.[3] In such cases, the axioms of the theory are formulated by hypothesis, and not by inductive inference.

Duhem remarked that scientific procedure is impregnated throughout with theoretical considerations. He supported Whewell's contention that there are no irreducible facts devoid of all theory. Duhem stressed that the scientist invariably interprets experimental findings with the aid of some theory. What is of interest to the scientist is not simply that the pointer of some instrument is on 3.5. Such an observation is of value only in conjunction with an interpretation of its meaning. For instance, the pointer reading is

---

* The root-mean-square velocity u is defined as follows:

$$u = \sqrt{\left(\frac{v_1^2 + v_2^2 + v_3^2 + \ldots v_n^2}{n}\right)}$$

where n is the number of molecules.

interpreted to mean that the current in a circuit is a certain value, that the temperature of a substance has a certain value, or something similar. Moreover, as Duhem pointed out, the scientist recognizes that the instruments he employs have a finite experimental error. For example, if a manometer is read '3.5', and if its limit of experimental error is ±0.1 atmosphere, then any pressure between 3.4 and 3.6 atmospheres is consistent with the reading. Duhem expressed this by suggesting that indefinitely many "theoretical facts" are consisent with a set of experimentally given conditions.[4]

On the basis of such considerations, Duhem criticized the ideal of scientific procedure which Newton had given in the *General Scholium* of the *Principia*. Newton had recommended that natural philosophy be restricted to propositions reached by inductive generalization from statements about phenomena. Even though Newton himself did not follow this inductivist ideal in the *Principia*, the ideal itself had proved tenacious in the history of science. Duhem observed that

two inevitable rocky reefs make the purely inductive course impracticable for the physicist. In the first place, no experimental law can serve the theorist before it has undergone an interpretation transforming it into a symbolic law; and this interpretation implies adherence to a whole set of theories. In the second place, no experimental law is exact, but only approximate, and is therefore susceptible to an infinity of distinct symbolic translations; and among all these translations, the physicist has to choose one which will provide him with a fruitful hypothesis, without his choice being guided by experiment at all.[5]

Campbell on "Hypotheses" and "Dictionaries"

N. R. Campbell, writing in 1919, made the distinction between an axiom system and its application to experience the basis of a careful analysis of the structure of physical theories. According to Campbell, a physical theory comprises statements of two different kinds. He termed one set of statements the "hypothesis" of the theory. In Campbell's usage, a "hypothesis" is a collection of statements the truth of which cannot be ascertained empirically.[6] It makes no sense to ask about the empirical truth of a hypothesis in itself, because no empirical meaning has been assigned to its terms.

Campbell included within the hypothesis of a theory both the axioms and the theorems deducible from them.

Campbell referred to the second set of statements within a theory as a "dictionary" for the hypothesis. Statements in the dictionary relate the terms of the hypothesis to statements whose empirical truth can be determined. Campbell's view of the structure of a scientific theory may be represented as follows:

In this diagram, α, β, γ, ... are the terms of the axiom system, and the lines joining the terms represent the axioms. In itself, the axiom system is a set of abstract relations among uninterpreted terms. The boundary between the axiom system and the realm of sense experience is bridged by dictionary entries which link certain terms of the axiom system with experimentally measurable properties.

In agreement with Duhem, Campbell emphasized that in many theories there are terms for which there are no dictionary entries. It is not necessary to link every hypothetical term to experimentally testable assertions in order to achieve empirical significance for a theory as a whole. In the diagram above, δ and ω are not mentioned in the dictionary. However, the entire axiom system, within which δ and ω are terms, is linked to experience through dictionary entries relating α and A, β and B, and γ and C.

The kinetic theory of gases is a good illustration of this point. The axioms of the theory state relations among the masses and velocities of individual molecules. But there is no dictionary entry for individual molecular velocities. Nevertheless, individual molecular velocities are related to the root-mean-square velocity of all the molecules, and the root-mean-square velocity is correlated through the dictionary with the temperature and pressure of the gas.

Mathematical Theories and Mechanical Theories

Campbell subdivided physical theories into "mathematical theories" and "mechanical theories", and based the subdivision on a difference in formal structure. Each important term of the hypothesis of a mathematical theory is correlated directly and separately with empirically determined magnitudes. Physical geometry exemplifies this type of theory. Terms such as 'point', 'line', and 'angle' are linked directly to measuring procedures. In the case of a mechanical theory, on the other hand, some of the terms of the hypothesis are correlated with empirically determined magnitudes only through functions of these terms.[7] This is the case for individual molecular velocities in kinetic theory. The kinetic theory of gases thus exemplifies the mechanical type of physical theory.

Analogies

Campbell held that the formal structure of a scientific theory consists of a hypothesis and a dictionary. But he also held that it is not sufficient for a theory merely to display the required formal structure. It must, in addition, be associated with an analogy. An acceptable theory exhibits an analogy to a system governed by previously established laws. And these previously established laws are judged to be more familiar, or more adequate, than the laws deduced from the theory. Campbell declared that a theory

always explains laws by showing that if we imagine that the system to which those laws apply consists in some way of other systems to which some other known laws apply, then the laws can be deduced from the theory.[8]

In the kinetic theory of gases, for instance, an analogy is drawn between the molecules of a gas and a swarm of particles. The particles are presumed to obey Newton's laws and to undergo collisions without loss of energy. This analogy played an important role in the historical development of theories about the behaviour of gases. Initially, the positive analogy between particles and molecules was restricted to the properties of motion and elastic impact. No reference was made to other properties that the particles may have. Subsequently, van der Waals extended the theory to account for the behaviour of gases under high pressures. He accomplished this by making certain assumptions about the volume of a particle and the forces existing between particles. These properties initially were part of the neutral analogy between particles and molecules.

Duhem and Campbell both were aware of the heuristic role of analogy in this instance. Bur for Duhem, to assert a theory is to assert a positive analogy only, whereas for Campbell, to assert a theory is to assert a positive-plus-neutral analogy. For this reason, Duhem described the transition from the original kinetic theory to its modification by van der Waals as the *replacement* of one theory by another, whereas Campbell described the transition as an *extension* of kinetic theory.

Campbell emphasized that the analogy associated with a theory is not merely a heuristic device to facilitate the search for additional laws. On the contrary, the analogy is an essential part of a theory, because it is only in terms of the analogy that a theory can be said to explain a set of laws. Campbell illustrated this point by formulating the following *ad hoc* theory:

The hypothesis consists of the following mathematical propositions:

(1) $u, v, w, \ldots$ are independent variables.
(2) a is a constant for all values of these variables.
(3) b is a constant for all values of these variables.
(4) $c = d$, where c and d are dependent variables.

The dictionary consists of the following propositions:

(1) The assertion that $(c^2 + d^2)a = R$ where R is a positive and rational number, implies the assertion that the (electrical) resistance of some definite piece of pure metal is R.

(2) The assertion that $\dfrac{cd}{b} = T$ implies that the (absolute) temperature of the same piece of pure metal is T[9].

It may be deduced from the hypothesis that

$$(c^2 + d^2)\, a = 2ab \left(\frac{cd}{b} \right).$$

According to the dictionary, this theorem is equivalent to the experimental law that the electrical resistance of the piece of pure metal is directly proportional to its absolute temperature.

What is wrong with such a theory? Duhem would say that it fails to achieve economy of representation, and that it is unlikely to have heuristic value. Campbell insisted, however, that this hypothesis-plus-dictionary is not a "theory" at all. The hypothesis and the dictionary have been formulated solely to imply the desired experimental law. But clearly, a particular law, or even a set of laws, may be deduced from indefinitely many sets of premisses. The successful deduction of a law from an hypothesis-plus-dictionary is a necessary, but not a sufficient, condition for explaining the law. According to Campbell, it is only when an analogy is drawn to other known laws that a theory explains the laws deducible from it.

Campbell believed this to be true for mathematical theories as well as for mechanical theories. But whereas the analogy for a mechanical theory is explicitly stated and obvious, such is not the case for a mathematical theory. Campbell explained this by pointing out that in a mathematical theory the laws to which the analogy is drawn are the same laws which are deduced from the theory. The analogy is one of mathematical form. The theory from which the experimental laws are deduced is of the same mathematical form as the laws themselves.

Campbell cited Fourier's theory of heat conduction as an example of a mathematical theory. This theory consists of a mathematical equation and a dictionary. The equation is

$$\lambda \left(\frac{\partial^2 \theta}{\partial x^2} + \frac{\partial^2 \theta}{\partial y^2} + \frac{\partial^2 \theta}{\partial z^2} \right) = \rho c \, \frac{\partial \theta}{\partial t}$$

The dictionary stipulates that θ is the absolute temperatrue, λ the thermal conductivity, ρ the density, c the specific heat, t the time, and x, y, z the spatial co-ordinates of a point in an infinitely long slab of material. Numerous experimental laws about the conduction of heat through finite slabs of various materials can be deduced from this theory. The experimental laws state relations among the same variables and constants as are mentioned in the theory, and the laws share with the theory a common mathematical form. According to Campbell, it is in virtue of this analogy between

Fourier's theory and the experimental laws of heat conduction that the theory may be said to explain the laws.

Campbell maintained that the aim of science is the discovery and explanation of laws, and that laws can be explained only by their incorporation in theories. His incisive analysis of the structure of scientific theories was a further blow against inductivist views of scientific procedure.

Mechanical theories, in particular, arise only upon successful application of an analogy. And no rules can be specified in advance to separate appropriate from inappropriate analogies. The imagination of the theorist is restricted only by the requirements of internal consistency and the deducibility of experimental laws. Once formulated, the mark of a successful mechanical theory is its fruitfulness in suggesting further correlations.

Mathematical theories also arise only upon the successful application of analogies. In this process, considerations of mathematical simplicity are important. But Campbell insisted that the formulation of a mathematical theory is not simply an extrapolation of experimental laws. The theorist must select among alternative mathematical relations which both imply the laws and exhibit some similarity of mathematical form to the laws. There is nothing in the experimental laws themselves which forces him to select one particular alternative.[10]

Campbell's claim that it is only in virtue of an analogy that a scientific theory may be said to explain laws deducible from it has been challenged by Carl Hempel. Hempel argued that Campbell's *ad hoc* theory about the electrical resistance of metals does not prove that an appeal to an analogy is essential to scientific explanation.

Hempel suggested a different *ad hoc* theory from which the law of resistance may be deduced. The hypothesis consists of the following two relations:

$$(1)\ c(u) = \frac{k_1 a(u)}{b(u)}, \text{ and } (2)\ d(u) = \frac{k_2 b(u)}{a(u)},$$

where k_1 and k_2 are constants. The dictionary specifies that, for any piece of pure metal u, $c(u)$ is its electrical resistance and $d(u)$ is the reciprocal of its absolute temperature.[11]

It may be deduced from the above hypothesis that

$$c(u) = k_1 k_2 \frac{1}{d(u)}.$$

In terms of the dictionary, this relation stipulates that the electrical resistance of a piece of pure metal is directly proportional to its absolute temperature.

Hempel pointed out that his theory, unlike that of Campbell, does display an analogy to a previously established law. Each of the relations stated in the hypothesis is a formal analogue of Ohm's Law.* But the existence of this analogy adds no explanatory power to the theory. As Duhem had observed, the explanatory power of a theory derives from arguments in which experimental laws are deduced, and analogies are not involved in these arguments. Hempel emphasized that both his own theory and Campbell's alternative theory are deficient in explanatory power because there is just one single experimental law which can be deduced within each theory. Neither theory achieves conceptual integration by showing how a particular set of theoretical assumptions implies a number of different experimental laws. According to Hempel, it is this conceptual integration, which Duhem had called the "representative function", that constitutes the explanatory power of a scientific theory.

Hempel conceded that analogies often are of value in guiding further research. He did not dispute the fact that analogies have been influential in the historical development of the sciences. But he did maintain, with Duhem, that since analogies do not occur as premisses in the deduction of experimental laws, analogies are not part of the structure of scientific theories.

The most that has been established by Hempel's counter-case is that not every appeal to a similarity of form provides an explanation for a set of laws. This leaves unaffected Campbell's claim that the explanation of laws by a theory is achieved only by formulating an analogy to *some* system governed by previously established laws. Campbell presumably would agree that the reference to Ohm's Law does not establish a proper analogue, and that Hempel's hypothesis-plus-dictionary has no explanatory power. But Campbell

---

* $i = \dfrac{V}{R}$, where i is the current, V the potential difference, and R the resistance, in an electrical circuit.

is committed only to the position that if a theory does have explanatory power, then it exhibits an analogy to a system governed by previously established laws. A "theory" which displays an analogy but which does not have explanatory power is not a counter-case to this claim.

Hesse on the Scientific Use of Analogies

Mary Hesse has suggested that to use an analogy in science often is to claim that two types of relations hold between an analogue and the system to be explained. The first is similarity relations between the properties of the analogue and the properties of the system to be explained. The second is causal relations, or functional relations, which hold both for the analogue and for the system to be explained. For instance, an analogy between the properties of sound and the properties of light may be represented as follows:

Causal Relations	Properties of Sound	Properties of Light
laws of reflection, refraction, *et al.*	echoes loudness pitch propagated in air	reflection brightness colour propagated in 'ether' .....

similarity relations

This analogy may be used to make a twofold claim. The first claim is that the corresponding properties in each column are similar. The second claim is that there are causal relations of the same type that link the terms within each column. These would include laws of reflection, refraction, the variation of intensity with distance, and the like. Hesse pointed out that each of these claims may be challenged. One may argue that the similarity relations are superficial. And one may argue that it is inappropriate to apply the known causal relations of sound propagation to the case of light propagation.[12]

The analogy used in Hempel's counter-case differs in an important respect from the sound–light analogy. In the sound–light analogy, horizontal similarity relations are presumed to hold independently of the existence of vertical causal relations. This is not the case in Hempel's analogy. The only relation claimed to hold between terms of the analogue and terms of the system to be explained is participation in functional relationships of the same form. Horizontal relatedness is established only in virtue of an identity of form in the respective vertical relations, viz.:

Functional relations	Properties of electrical circuits	Properties of a piece of pure metal			
		Axiom (1)		Axiom (2)	
① ∝ ②/③	① i	①	$c(u)$	①	$d(u)$
	② V	②	$a(u)$	②	$b(u)$
	③ R	③	$b(u)$	③	$a(u)$

Hesse referred to analogies of this type as "formal analogies" to distinguish them from "material analogies" which do have horizontal similarity relations that are independent of vertical relations.[13]

Hesse maintained that the acceptability of formal analogies depends entirely on the appropriateness of the formal relations cited. In Hempel's counter-case, there would seem to be no reason (apart from establishing a deductive relation yielding the known law) to select Ohm's Law as analogue. For the purpose of deducing the known law, the Ideal Gas Law* would be an equally good analogue. We have been given no reason to believe that there is any connection between Hempel's axioms and the flow of current in an electrical circuit. What is needed at this point is a criterion of appropriateness of analogical links.

Harré on the Importance of Underlying Mechanisms

In opposition to the Duhem–Hempel view of theories, Rom Harré has recommended a "Copernican Revolution" in which emphasis is shifted from the formal, deductive structure of theories to the associated models. He has declared that

---

* $P = k \dfrac{T}{V}$, which also has the form ① ∝ ②/③.

the Copernican revolution in the philosophy of science consists in bringing models into the central position as instruments of thought, and relegating deductively organized structures of propositions to a heuristic role only, and resurrecting the notion of the generation of one event or state of affairs by another. On this view theory construction becomes essentially the building up of ideas of hypothetical mechanisms.[14]

Harré maintained that this emphasis is more consistent with "the persistent intuitions of scientists"[15] than is the position of Duhem.

Harré distinguished three component parts of a scientific theory: statements about a model, empirical laws, and transformation rules. The statements about a model typically include both hypotheses that assert the existence of theoretical entities and hypotheses about the behaviour of these entities. The transformation rules may comprise both causal hypotheses and modal transforms. Causal hypotheses may be expressed in conditional sentences of the form 'If M then E', where 'M' is a state of the model, and 'E' is a type of observed effect. Modal transforms may be expressed in bi-conditional sentences of the form 'M if, and only if, E'.

On this analysis, the structure of the kinetic theory of gases would be represented, in part, as follows:

Model	Transformation Rules	Empirical Laws
Existential hypotheses 'There exist molecules.'	Causal 'Pressure is caused by molecular impacts.' ('If I then P.')	$\dfrac{PV}{T}$ = constant
Descriptive hypotheses 'Collisions are elastic.' '$\Delta m_i v_i$ = constant.' . . .	Modal 'Temperature is the mean kinetic energy of the molecules.' ('T if, and only if, $\dfrac{2}{3}\dfrac{E}{k}$')	. . .

With respect to the model embedded in theories, Harré emphasized the existential hypothesis suggested by the model rather than the deductive structure which may be developed from the descriptive hypotheses. He insisted that the formulation of existential hypotheses is a "science-extending" operation, and supported this contention by analyses of the historical development of science. It is incontestable that attempts to justify claims about the existence of theoretical entities such as capillaries, radio waves, and neutrinos, have contributed to scientific progress.

Harré indicated the spectrum of possible outcomes of attempts to confirm existential hypotheses. One possibility is that both the demonstrative and the recognitive criteria for the type of entity sought are satisfied. Mendeleef's predictions of the existence of hitherto undiscovered elements is an example. The recognitive criteria which he specified—physical properties, types of compounds formed, *et al.*—subsequently were shown to be satisfied by Scandium, Gallium, and Germanium. Much the same may be said for hypotheses about the existence of positrons, viruses, and neutrinos.

In other cases, existential hypotheses may be abandoned because demonstrative criteria have not been met. This was the fate of the hypothesis that there exists a planet whose orbit is inside that of Mercury, and also the hypothesis that there exists an ether within which light is propagated.

And in still other cases, existential hypotheses may be abandoned because recognitive criteria have not been met. In such cases, the demonstration-region is found to be occupied by something that does not satisfy the original recognitive criteria. For instance, microscopic investigations of the human heart revealed that it is a continuous muscle, and Galen's hypothesis that there exist pores in the septum through which blood passes, was abandoned.

In some instances, failure to meet recognitive criteria has resulted in recategorization of the theoretical entity in question. This happened in the case of "caloric". Many eighteenth-century scientists explained thermal effects in terms of the transfer of an invisible fluid. But in the nineteenth century, various studies indicated that caloric did not satisfy certain recognitive criteria that should be met by substantive entities. For example, this "substance" largely disappeared in certain processes in which mechanical work is performed. One response of scientists was to reinterpret caloric as a quality of substance—the average kinetic energy of its constituent particles—rather than as a substance itself.

According to Harré, one criterion of the appropriateness of analogical links embedded in a theory is the generation of existential hypotheses from the theory. If no existential hypotheses are suggested by a theory, then the theory does not advance our understanding of the underlying mechanisms of natural processes. Harré declared that

scientific explanation consists in finding or imagining plausible generative mechanisms for the patterns amongst events, for the structures of things, for the generation, growth, decay, or extinction of things and materials, for changes within persisting things and materials.[16]

From this standpoint, the theories which Campbell and Hempel formulated to deduce the variation of electrical resistance with temperature, are wholly inadequate.

Notes

[1] Hermann von Helmholtz, 'On the Origin and Significance of Geometrical Axioms', trans. E. Atkinson, in *Helmholtz: Popular Scientific Lectures*, ed. M. Kline (New York: Dover Publications, 1962), 239–47.

[2] Pierre Duhem, *The Aim and Structure of Physical Theory*, trans. P. Wiener (New York: Atheneum, 1962), 32.

[3] Ibid. 207.

[4] Ibid. 135–6.

[5] Ibid. 199.

[6] N. R. Campbell, *Foundations of Science* (New York: Dover Publications, 1957), 122.

[7] Ibid. 150.

[8] Campbell, *What Is Science?* (New York: Dover Publications, 1952), 96.

[9] Campbell, *Foundations*, 123.

[10] Ibid. 153.

[11] Carl Hempel, *Aspects of Scientific Explanation and Other Essays in the Philosophy of Science* (New York: Free Press, 1965), 444.

[12] Mary Hesse, *Models and Analogies in Science* (Notre Dame, Ind.: University of Notre Dame Press, 1966), 80–1.

[13] Ibid. 68–9.

[14] Rom Harré, *The Principles of Scientific Thinking* (London: Macmillan, 1970), 116.

[15] Ibid. 116.

[16] Ibid. 125.

10

Inductivism *v.* the Hypothetico-Deductive View of Science

JOHN STUART MILL (1806–73) received intensive instruction from his father James Mill, a respected economist, historian, and philosopher. The instruction ranged from Greek, commenced at the age of three, to psychology and economic theory. Mill was associated with the East India Company (1823–58), and was elected to Parliament in 1865, where he worked for woman's suffrage and the reform of land tenure in Ireland. He published numerous books and essays in support of the philosophy of utilitarianism.

The elder Mill impressed upon his son the importance of collecting and weighing evidence, and John Stuart sought to formulate inductive techniques for assessing the connection between conclusions and evidence. He discovered that implicit in the methodology of the sciences are rules of proof of causal connection. Mill set forth his philosophy of science in *System of Logic* (1843) in which he acknowledged his debt to Herschel and Whewell.

WILLIAM STANLEY JEVONS (1832–82) was appointed Professor of Logic and Political Economy at the University of Manchester in 1866 and subsequently taught at University College, London. He made contributions to logic and the theory of probability, and pioneered the application of statistical methods in meteorology and economics. Jevons opposed Mill's inductivism on behalf of a hypothetico-deductive view of science in the tradition of Whewell.

Mill's Inductivism

Inductivism is a point of view that emphasizes the importance to science of inductive arguments. In its most inclusive form, it is a thesis about both the context of discovery and the context of justification. With respect to the context of discovery, the inductivist position is that scientific inquiry is a matter of inductive generalization from the results of observations and experiments. With respect to the context of justification, the inductivist position is that a scientific law or theory is justified only if the evidence in its favour conforms to inductive schemata.

John Stuart Mill's philosophy of science is an example of the inductivist point of view. Mill made certain extreme claims about the role of inductive arguments both in the discovery of scientific laws and in the subsequent justification of these laws.

Context of Discovery

Mill's Inductive Methods. Mill was an effective propagandist on behalf of certain inductive methods which had been discussed by Duns Scotus, Ockham, Hume, and Herschel, among others. So much so that these methods came to be known as "Mill's Methods" of experimental inquiry. Mill stressed the importance of these methods in the discovery of scientific laws. Indeed, in the course of a debate with Whewell, Mill went so far as to claim that every causal law known to science had been discovered "by processes reducible to one or other of those methods".[1]

Mill discussed four inductive methods.* They may be represented as follows:

---

* Mill also discussed a fifth method, a Joint Method of Agreement and Difference, in which these two methods are combined in a single schema.

Agreement

Instance	Antecedent circumstances	Phenomena
1	ABEF	abe
2	ACD	acd
3	ABCE	afg

Therefore A is the cause of a.

Difference

Instance	Antecedent circumstances	Phenomena
1	ABC	a
2	BC	—

Therefore A is an indispensable part of the cause of a.

Concomitant variations

Instance	Antecedent circumstances	Phenomena
1	$A^+ BC$	a^+b
2	$A^0 BC$	a^0b
3	$A^- BC$	a^-b

Therefore A and a are causally related.

Residues

Antecedent circumstances	Phenomena
ABC	abc
B is the cause of	b
C is the cause of	c

Therefore A is the cause of a.

Mill maintained that the Method of Difference is the most important of the four methods. In his summary statement of this schema, he observed that circumstance A and phenomenon a are causally related only if the two instances differ in one, and only one, circumstance.[2] But if this restriction were enforced, no causal relation could be uncovered by application of the Method of Difference.

The description of two instances involves reference either to different places or to different times, or both. But since there is no reason *a priori* to exclude from the list of circumstances position in space and time, it is not possible that two instances which differ with respect to the occurrence of a phenomenon, differ also in one circumstance only.

A further difficulty is that, in Mill's summary statement of the method, all circumstances are on a par. To explain, for instance, why nitroglycerin exploded on one occasion and not on another, one would have to specify, not only the ways in which the substance was handled, but also the number of clouds in the sky and the extent of sunspot activity. If all circumstances were on a par, one could specify an instance adequately only by describing the state of the entire universe at a particular time.

Mill was aware of this. He conceded that the usefulness of Difference as a method of discovery depends on the assumption that, for any particular inquiry, only a small number of circumstances need be considered. However, he maintained that this assumption itself is justified by experience. Mill claimed that, for a great number of cases, the schema of the Method of Difference is satisfied, even though the inquiry is restricted to a small number of circumstances.

This may be so. But then the discovery of causal relations involves more than a mere specification of values which fit the schema. In order to use this method in scientific inquiry, a hypothesis must be made about which circumstances *could be* relevant to the occurrence of a given phenomenon. And this hypothesis about relevant circumstances must be formulated prior to application of the schema. Hence Mill's claim that application of the Method of Difference is sufficient to uncover causal relations must be rejected. On the other hand, once a supposition has been made that a circumstance is related to a phenomenon, the Method of Difference specifies a valuable technique for testing the supposition by means of controlled experiments.

Mill regarded the Method of Difference to be the most important instrument for discovering causal relations. His claims on behalf of the Method of Agreement were more modest. He maintained that the Method of Agreement is a useful instrument for the discovery of scientific laws. But he acknowledged that this method is subject to important limitations.

One limitation is that the method is effective in the search for causal relations only if an accurate inventory of relevant circumstances has been made. If a relevant circumstance present in each instance is overlooked, application of the Method of Agreement may mislead the investigator. Hence, successful applications of Agreement—like successful applications of Difference—are possible only on the basis of antecedent hypotheses about relevant circumstances.

An additional limitation of the Method of Agreement arises from the possibility that a plurality of causes is at work. Mill acknowledged that a particular type of phenomenon may be the effect of different circumstances on different occasions. In the schema above, for instance, it is possible that B caused a in instances 1 and 3, and that D caused a in instance 2. Because this possibility exists, one may conclude only that it is *probable* that A is the cause of a. Mill noted that it is a function of the theory of probability to estimate the likelihood that a plurality of causes is present, and he pointed out that, for a given correlation, this probability may be decreased by including additional instances in which the circumstances are further varied and yet the correlation remains.

Mill believed that the possibility of a plurality of causes can cast no doubt on the truth of conclusions reached by the Method of Difference. He declared that, for any particular argument by Difference

it is certain that in this instance at least, A was either the cause of a, or an indispensable portion of its cause, even though the cause which produces it in other instances may be altogether different.[3]

But what does it mean to speak of "a cause in this instance"? Mill previously had defined a cause to be a circumstance, or set of circumstances, both invariably and unconditionally followed by an effect of a given type. It would seem that Mill's position in the quotation cited above is that a single application of the Method of Difference can establish that each occurrence of a circumstance must be followed by a corresponding phenomenon. Presumably this is the case in spite of the acknowledged possibility that some other set of circumstances also may be followed by the phenomenon in question. This conclusion about Mill's meaning may be supported by citing Mill's claim that a

plurality of causes ... not only does not diminish the reliance due to the Method of Difference, but does not even render a greater number of observations or experiments necessary: two instances, the one positive and the other negative, are still sufficient for the most complete and rigorous induction.[4]

W. S. Jevons subsequently pointed out that Mill had made an unjustified leap from a statement about what takes place in a single experiment to a generalization that what takes place in one experiment also will take place in other experiments.[5]

Multiple Causation and the Hypothetico-Deductive Method. It is common practice in historical studies of the philosophy of science to contrast the views of Mill and Whewell. Often Mill is presented as identifying scientific discovery with the application of inductive schemata, whereas Whewell is presented as viewing scientific discovery as a free invention of hypotheses.

No doubt Mill did make incautious claims for his inductive methods. The methods certainly are not the sole instruments of discovery in science. But despite the comments that Mill directed against Whewell on this issue, Mill clearly recognized the value of hypothesis—formation in science. It is unfortunate that subsequent writers have overemphasized the incautious claims that Mill made in his debate with Whewell.

In a discussion of multiple causation, for example, Mill greatly restricted the range of applicability of his inductive methods. Instances of multiple causation are instances in which more than one cause is involved in the production of an effect. Mill subdivided cases of multiple causation into two classes: instances in which the various causes continue to produce their own separate effects, and instances in which there is a resultant effect other than the effects that would be produced separately. Mill further subdivided the latter class into instances in which the resultant effect is the "vectorial sum" of the causes present, and instances in which the resultant effect differs in kind from the several effects of the separate causes.

Mill held that the "Mutual Coexistence of Separate Effects" may be analysed successfully by the four inductive methods. In addition, he held that the same is true of "Resultant Effects Different in Kind". He noted that in this latter type of situation the investigator may correlate the effect with the presence or absence of circumstances, and then apply the Methods of Agreement and Difference.

Mill believed the situation to be quite different in the case of the

Mill's View of Multiple Causation

"Composition of Causes". This type of multiple causation is not amenable to investigation by the four inductive methods. Mill cited the case of motion caused by two impressed forces. The result is motion along the diagonal of a parallelogram, the sides of which have lengths proportional to the magnitudes of the forces.

The Parallelogram of Forces

There is no question here of conjoined causes giving rise to an effect different in kind from the separate effects of the respective causes. Each separate component cause is fulfilled, but fulfilled in such a way as to produce a reinforcement or cancellation of effects. This is true even in dynamic equilibrium, where the net effect of the forces acting is rest.

An important consideration about the composition of forces is that the contribution of the several forces acting cannot be determined from information about the resultant motion. There are indefinitely many sets of forces which could produce a given resultant motion.

Mill concluded that his inductive methods were unavailing in cases of the Composition of Causes—one cannot proceed inductively from knowledge that a resultant effect has occurred to knowledge of its component causes. For this reason, he recommended that a "Deductive Method" be employed in the investigation of Composite Causation.

Mill outlined a three-stage Deductive Method: (1) the formulation of a set of laws; (2) the deduction of a statement of the resultant effect from a particular combination of these laws; and (3) verification. Mill preferred that each law be induced from a study of the relevant cause acting separately, but he allowed the use of hypotheses not induced from phenomena. Hypotheses are suppositions about causes which may be entertained by a scientist in cases where it is not practical to induce the separate laws.

Mill agreed with Whewell that the use of hypotheses is justified if their deductive consequences agree with observations. However, Mill set very severe requirements for the full verification of hypotheses. He demanded of a verified hypothesis, not only that its deductive consequences agree with observations, but also that no other hypothesis imply the facts to be explained. Mill maintained that the complete verification of a hypothesis requires the exclusion of every possible alternative hypothesis.

Mill held that complete verification is achieved sometimes in science, but he cited just one example—Newton's hypothesis of an inverse-square central force between the sun and the planets. Mill claimed that Newton had shown, not only that the deductive consequences of this hypothesis were in agreement with the observed motions of the planets, but also that no other force law could account for these motions.[6] But neither Mill nor Newton advanced a proof that the alternatives examined exhaust the possible ways of accounting for the motions of the planets.

Mill believed that this was a case of multiple causation in which complete verification had been achieved. However, he was aware of the difficulty of excluding alternative hypotheses, and, in other cases, he was most cautious in assessing the status of hypotheses and theories. He maintained, for instance, that although the wave theory of Young and Fresnel had many confirmed deductive consequences, such confirmation was not tantamount to verification. Mill suggested that, at some future time, a theory may be formulated which explains not only the phenomena in his day explained by the wave theory, but also those absorption and emission phenomena not explained by the theory.[7] Consistent with the stringent requirements of his concept of verification, Mill maintained an admirably open-minded attitude towards the theories of his day.

Mill attributed to the Deductive Method an important role in scientific discovery. He declared that to it

the human mind is indebted for its most conspicuous triumphs in the investigation of nature. To it we owe all the theories by which vast and complicated phenomena are embraced under a few simple laws, which, considered as the laws of those great phenomena, could never have been detected by their direct study.[8]

On this point Mill and Whewell were in agreement. Both were convinced that the great Newtonian synthesis was the fruit of a hypothetico-deductive method. This being the case, one must conclude that Mill did not defend an exclusively inductivist position about the context of scientific discovery.

Context of Justification

Although Mill did not reduce scientific inquiry to the application of inductive schemata, he did insist that the *justification* of scientific laws is a matter of satisfying inductive schemata. He held that it is the function of inductive logic to provide rules for the appraisal of judgements about causal connection. According to Mill, a statement about a causal connection may be justified by showing that the evidence in its favour conforms to specific inductive schemata.

Causal Relations and Accidental Relations. Mill maintained that an important goal of science is proof of causal connections. He based his discussion of this goal on an analysis of Hume's position that causal relations are nothing but constant sequential conjunctions of two types of events. Mill recognized that if Hume were correct to equate causal relation and constant conjunction, then all invariable sequences would be on a par. But according to Mill, some invariable sequences are causal and some are not. For instance, the addition of a lump of sodium to a glass of water is the cause of the vigorous production of bubbles in the water. But day is not the cause of night, despite the fact that our experience to date has revealed this sequence to be invariable. Mill therefore distinguished causal sequences from accidental sequences. He insisted that a causal relation is a sequence of events which is *both invariable and unconditional*, thereby allowing for the possibility that some invariable sequences are non-causal.

Mill acknowledged that the distinction between causal and non-causal sequences is of value only if some way can be found to establish that some sequences are unconditional. He suggested that an unconditional sequence is a sequence which not only has been invariable in our past experience, but also will continue to be so, "as long as the present constitution of things endures".[9] He explained that he meant by "the present constitution of things" those "ultimate laws of nature (whatever they may be) as distinguished from the derivative laws and from the collocations".[10]

Mill suggested that the status of an invariable sequence may be decided by considering what would happen if the conditions within which the sequence ordinarily takes place are altered. If these conditions can be altered in a way which is consistent with the "ultimate laws", and if the effect then would fail to occur, then the sequence is a conditioned sequence. In the case of day and night, for example, Mill noted that the relevant conditions of this sequence include the diurnal rotation of the Earth, radiation from the sun, and the absence of intervening opaque bodies. He maintained that, since the failure of any one of these conditions to hold would not violate the ultimate laws of nature, the sequence day—night is a conditioned sequence.

The general usefulness of this approach is severely limited by Mill's failure to specify which laws are the "ultimate laws of nature". Mill did not pursue this approach further. He remained convinced, however, that causal sequences do differ from accidental sequences, and that this difference can be exhibited within experience. What is needed, Mill believed, is a theory of proof which stipulates the form of valid inductive arguments. Such a theory would enable a philosopher of science to determine which generalizations from experience state causal relations.

Upon occasion, Mill extolled all four of his inductive schemata as rules of proof of causal connection. In his more cautious moments, however, he restricted the proof of causal connection to those arguments which satisfy the Method of Difference.

Justification of Induction. In order to establish that any argument which has the form of the Method of Difference proves causal connection, Mill would have to show that the connection is both invariable and unconditional. Mill believed that he could do this. However, philosophers of science are in general agreement that Mill failed to prove his case. Mill's arguments to substantiate his claim are based on two premisses, and he failed to establish that either premiss is true.

The first premiss is that the positive and negative instances which fit the schema of Difference differ in just one relevant circumstance. But as noted above, Mill could not establish this. The best he could do was to show that in many cases sequences have been observed to be invariable despite the fact that only a small number of circumstances have been taken into account. But this does not suffice to prove that no further circumstance could be relevant to the occurrence and non-occurrence of the phenomenon.

The second premiss is a principle of universal causation, which stipulates that for every phenomenon there is some one set of antecedent circumstances upon which it is invariably and unconditionally consequent. Mill demanded that the truth of the law of causation be established on empirical grounds, and he acknowledged that, in this demand, he was confronted by a paradox. The paradox is that, if the law of causation is to be proved by experience, then it must be itself the conclusion of an inductive argument. But every inductive argument that proves its conclusion presupposes the truth of the law of causation. Mill conceded that his proof appeared to involve a vicious circle. He recognized that he could not prove the law of causation by an inductive argument using the Method of Difference. To do so would be circular, since the law of causation is needed to justify the Method of Difference itself.

Mill thought that he could avoid closing the circle by means of a thesis about inductive arguments by simple enumeration. He maintained that

the precariousness of the method of simple enumeration is in an inverse ratio to the largeness of the generalization. The process is delusive and insufficient, exactly in proportion as the subject-matter of the observation is special and limited in extent. As the sphere widens, this unscientific method becomes less and less liable to mislead; and the most universal class of truths, the law of causation for instance ... [is] duly and satisfactorily proved by that method alone.[11]

Thus, whereas the generalization 'all ravens are black' is precarious (remember the discovery of black swans), the generalization 'for every event of a given type there is a set of circumstances upon which it is invariably and unconditionally consequent' is not.

Mill held that the law of causation is a generalization of such breadth that every sequence of events affords a test of its truth. He also held that we do not know a single exception to this law. According to Mill, every seeming exception "sufficiently open to our observation", has been traced either to the absence of an antecedent circumstance ordinarily present, or to the presence of a circumstance ordinarily absent.[12] He concluded that, because every sequence of events is a test of the law of causation, and because every sequence investigated has confirmed the law, the law itself is a necessary truth.

Mill thus claimed to have demonstrated that an inductive argument by simple enumeration from empirical premises proves the law of causation to be a necessary truth. However, Mill's "proof" is not successful. No appeal to experience, to the way things are, proves that things could not be otherwise. Even if Mill could make good on his claim that there never has been a bona fide exception to the law of causation, this would not prove the law to be a necessary truth. And Mill requires that the law of causation be a necessary truth in order to justify his claim that arguments which fit the Method of Difference *prove* causal connections.

Jevons' Hypothetico-Deductive View

Mill's inductivist thesis about the context of justification was challenged at once by Jevons. Jevons insisted that to justify a hypothesis one must do two things. One must show that it is not inconsistent with other well-confirmed laws. And one must show that its consequences agree with what is observed.[13] But to show that a hypothesis has consequences that agree with what is observed is to utilize *deductive* arguments. Jevons thus rejected Mill's claim that the justification of hypotheses is by satisfaction of inductive schemata. In so doing, Jevons reaffirmed the emphasis placed on deductive testing by Aristotle, Galileo, Newton, Herschel, and many others.

Notes

[1] J. S. Mill, *System of Logic* (London: Longmans, Green, 1865), i. 480.

[2] Ibid. i. 431.

[3] Ibid. i. 486.

[4] Ibid. i. 485.

[5] W. S. Jevons, *Pure Logic and Other Minor Works* (London: Macmillan, 1890), 295.

[6] Mill, *System of Logic*, ii. 11–13.

[7] Ibid. ii. 22.

[8] Ibid. i. 518.

[9] Ibid. i. 378.

[10] Ibid. i. 378n.

[11] Ibid. ii. 101.

[12] Ibid. ii. 103.

[13] Jevons, *The Principles of Science* (New York: Dover Publications, 1958), 510–11.

11

Mathematical Positivism and Conventionalism

GEORGE BERKELEY (1685–1753) was born in Ireland of English stock. He was educated and later taught at Trinity College, Dublin. A devout Anglican, Berkeley was appointed Dean of Derry in 1724. Shortly thereafter, he sought to found a college in Bermuda, a project which failed for lack of funds. He assumed duties as Bishop of Cloyne in 1734. Berkeley's anti-materialist philosophy is set forth in the *Treatise Concerning the Principles of Human Knowledge* (1710) and *Three Dialogues Between Hylas and Philonous* (1713). His later writings include a critique of Newton's version of the differential calculus (*The Analyst*, 1734), and a positivistic critique of Newton's physics (*De Motu*, 1721).

ERNST MACH (1838–1916) was a Vienna-educated physicist who made contributions to mechanics, acoustics, thermodynamics, and experimental psychology, in addition to the philosophy of science. He crusaded against the intrusion of "metaphysical" interpretations into physics. Against the view that science should seek to describe some "objective reality"—e.g. atoms—behind appearances, Mach insisted

:at science should aim at an economical description of the relations among phenomena.

HENRI POINCARÉ (1854–1912) was born at Nancy into a distinguished family. His cousin Raymond was President of the French Republic during World War I. Poincaré attended the École des Mines with the intention of becoming a mining engineer, but his interests shifted to pure and applied mathematics. After a brief period at the University of Caen, he joined the faculty of the University of Paris (1881). Poincaré made important contributions to pure mathematics and celestial mechanics. His 1906 paper on the electron anticipated some of the results achieved by Einstein in the Special Theory of Relativity. Poincaré's writings on the philosophy of science—*Science and Hypothesis* (1905) and *The Value of Science* (1907)—emphasized the role of conventions in the formulation of scientific theories.

KARL POPPER (1902–   ) was Professor of Logic and Scientific Method at the University of London. In the influential *Logic of Scientific Discovery* (German 1934, English 1959), Popper criticized the Vienna Circle's search for a criterion of empirically meaningful statements, and suggested instead that empirical science be demarcated from pseudo-science with respect to methodology practised. He has reaffirmed and augmented this position in *Conjectures and Refutations* (1963). During World War II, Popper published *The Open Society and its Enemies*, an attack on Plato, Hegel, Marx, and all thinkers who would impose inexorable laws on history.

Berkeley's Mathematical Positivism

One of the early critics of Newton's philosophy of science was George Berkeley, a philosopher who achieved a measure of notoriety for having advanced a number of arguments to prove that "material substances" do not exist. In his criticism of Newton, Berkeley accused Newton of failing to heed his own warnings. Newton had warned that it was one thing to formulate mathematical correlations involving forces, and quite another thing to discover what forces are "in themselves". Berkeley held that Newton was correct to distinguish his mathematical theories of refraction and gravitation from any hypotheses about the "real nature" of light and gravity. What distressed Berkeley was that Newton, under the guise of suggesting "queries", did talk about forces as if they were something more than terms in equations. Berkeley maintained that "forces" in mechanics were analogous to epicycles in astronomy. These mathematical constructions are useful in calculating the motions of bodies. But according to Berkeley, it is a mistake to attribute to these constructions a real existence in the world.

Berkeley maintained that the entire content of Newtonian mechanics is given in a set of equations, together with the claim that bodies do not move themselves. Berkeley was quite willing to grant Newton's claim that bodies do not have the power of self-movement. But he cautioned that Newton's references to "attractive forces", "cohesive forces", and "dissolutive forces" are apt to mislead the reader. These "forces" are mathematical entities only. Berkeley declared that

mathematical entities have no stable essence in the nature of things; and they depend on the notion of the definer. Whence the same thing can be explained in different ways.[1]

Berkeley thus defended an instrumentalist view of the laws of mechanics. He held that these laws are nothing but computational devices for the description and prediction of phenomena. And he insisted that neither the terms that occur in the laws nor the functional dependencies expressed by the laws need refer to anything that exists in nature. Berkeley maintained, in particular, that we have no knowledge of any referents for such terms as 'attractive force', 'action', and 'impetus'. We know only that particular bodies move in certain ways under certain conditions. Nevertheless, Berkeley conceded that terms such as 'attractive force' and 'impetus' have an important use in mechanics, in virtue of their occurrence in theories which enable us to predict sequences of events.

Berkeley opposed that view of science which likens science to cartography. Scientific laws and theories are not like maps. Each entry on a topographical map designates a feature of the terrain. And the adequacy of a map's representation may be ascertained in a reasonably straightforward way. But it is not the case that each term of a scientific theory must designate an independently knowable object, property, or relation in the universe.

Berkeley's instrumentalist emphasis is consistent with, and perhaps is derived from, his metaphysical thesis that the universe contains only two kinds of entities—ideas and minds. His summary statement of this position is that "to be is to perceive or to be perceived". On this view, minds are the sole causal agents. Forces cannot be causally efficaceous.

Moreover, Berkeley urged, no distinction can be enforced between "primary qualities" which are objective properties of bodies, and "secondary qualities" which exist only in the perceptual experience of the subject. Galileo, Descartes, and Newton had accepted the distinction between primary and secondary qualities, and had suggested that extension, position, and motion were primary qualities. Berkeley, however, denied that there are any primary qualities of bodies. He insisted that extension and motion are sensible qualities quite on a par with heat and brightness. Any knowledge that we have about the extension and motion of bodies is given to us in our perceptual experience.

Berkeley held that it is meaningless to talk, as Newton had done, about motions in Absolute Space. Space is not something that exists apart from, and independently of, our perception of bodies. Berkeley pointed out that if there were no bodies in the universe, then there would be no possible way to assign spatial intervals. He concluded that if it is not possible to assign spatial intervals in this situation, then it is meaningless to speak of a "space" devoid of all bodies.

In addition, Berkeley pointed out that if every body save one were annihilated, then no motion could be assigned to this body. This is because all motion is relative. To speak of a body's motion is to speak of its changing relations to other bodies. The motion of a single body within an Absolute Space is inconceivable.

Nor does Newton's bucket experiment establish the existence of Absolute Space. Berkeley correctly observed that the motion of water in the bucket is not a "truly circular motion", since it is compounded, not only of the motion of the bucket, but also of the earth's rotation and revolution around the sun. He concluded that this motion which Newton had cited as rotation with respect to Absolute Space may be referred instead to bodies in the universe other than the bucket.[2]

In the application of his theory of mechanics, Newton was forced to substitute relative spatial intervals for distances in Absolute Space. Berkeley suggested that Newton's references to motions in Absolute Space could be eliminated from physics without in any way impoverishing the discipline. He maintained that, whereas 'attractive force' and 'impetus' are useful mathematical fictions,

'Absolute Space' is a useless fiction and should be eliminated from physics. He recommended that the fixed stars be taken as specifying a reference frame for the description of motions.

Mach's Reformulation of Mechanics

In the latter part of the nineteenth century, Ernst Mach developed a critique of Newton's philosophy of science that was strikingly similar to the critique given by Berkeley. Mach shared Berkeley's instrumentalist view of scientific laws and theories. He declared that

it is the object of science to replace, or *save*, experiences, by the reproduction and anticipation of facts in thought.[3]

According to Mach, scientific laws and theories are implicit summaries of facts. They enable us to describe and anticipate phenomena. A good example is Snel's law of refraction. Mach observed that, in nature, there are various instances of refraction, and that the law of refraction is a "compendious rule" for the mental reconstruction of these facts.[4]

Mach suggested a Principle of Economy as a regulative principle for the scientific enterprise. He stated that

science itself . . . may be regarded as a minimal problem, consisting of the completest possible presentment of facts with the *least possible expenditure of thought*.[5]

The scientist should seek to formulate relations that summarize great numbers of facts. Mach stressed that a particularly effective way of achieving economy of representation is the formulation of comprehensive theories in which empirical laws are deduced from a few general principles.

Mach also shared Berkeley's conviction that it is a mistake to assume that the concepts and relations of science correspond to that which exists in nature. He conceded, for instance, that theories about atoms may be useful for the description of certain phenomena, but he insisted that this provides no evidence for the existence of atoms in nature.

Like Berkeley, Mach refused to posit a realm of "reality"— whether of primary qualities, atoms, or electric charges—behind the realm of appearance. His phenomenalism was quite as thorough-going as that of Berkeley. Mach declared that

in the investigation of nature, we have to deal only with knowledge of the connexion of appearances with one another. What we represent to ourselves behind the appearances exists *only* in our understanding, and has for us only the value of a *memoria technica* or formula, whose form, because it is arbitrary and irrelevant, varies very easily with the standpoint of our culture.[6]

Mach sought to reformulate Newtonian mechanics from a phenomenalist standpoint. He hoped to show, by means of this reformulation, that mechanics may be divested of "metaphysical" speculations about motions in Absolute Space and Time. The re-formulation took the form of a subdivision of the fundamental propositions of mechanics into two classes—empirical gener-alizations and *a priori* definitions.

According to Mach, the basic empirical generalizations of mech-anics are (1) that

bodies set opposite each other induce in each other, under certain circum-stances to be specified by experimental physics, contrary accelerations in the direction of their line of junction;

(2) that the mass-ratio of two bodies is independent of the physical states of the bodies; and (3) that the accelerations which each body A, B, C, ... induces in body K are independent of each other.

To these empirical generalizations, Mach added definitions of 'mass-ratio' and 'force'. The 'mass-ratio' of two bodies is "the negative inverse ratio of the mutually induced accelerations of those bodies", and 'force' is the 'product of mass and acceleration'.[7]

Mach regarded the empirical generalizations as contingent truths which are confirmed by experimental evidence. Supposedly, these generalizations would be falsified if the results of experiments turn out to be different than hitherto observed.

Mach emphasized that the generalizations in his reformulation become empirically significant only upon specification of procedures for measuring spatial intervals and temporal intervals. He suggested that spatial intervals be measured relative to a co-ordinate system defined by the "fixed" stars, thereby eliminating all reference to Absolute Space. He also insisted that because it is meaningless to speak of a motion "uniform in itself", references to Absolute Time be eliminated. According to Mach, temporal intervals must be measured by physical processes.

But even if satisfactory physical procedures can be found for determining spatio-temporal intervals, it may be argued that Mach has not established that the empirical generalizations of his re-formulation are subject to the possibility of being falsified. The phrase 'under certain circumstances to be specified by experimental physics', which occurs in the first generalization, conceals a problem. The physicist seeks to test the generalization for isolated systems which are unaffected by changes external to the system itself. But failure to record "contrary accelerations in the direction of their line of junction" may be taken to prove, not that the generalization is false, but that the two bodies have been incompletely isolated from disturbing influences. A physicist interested in preserving at all costs the generalization in question could use it as a convention to determine whether a system of bodies qualifies as an isolated system. As a convention, this relation would be subject to neither confirmation nor refutation.

Duhem on the Logic of Disconfirmation

The conventionalist point of view received further support from Pierre Duhem's analysis of disconfirmation of hypotheses. Duhem emphasized that the prediction that a phenomenon will occur is made from a set of premises which include laws and statements about antecedent conditions.

Consider a case in which the law 'all blue litmus paper turns red in acid solution' is tested by placing a piece of paper in a liquid. We predict that the paper turns red on the basis of the following deductive argument:

> L For all cases, if a piece of blue litmus paper is placed in an acid solution, then it turns red.
>
> C A piece of blue litmus paper is placed in an acid solution.
> _____
>
> $\therefore E$ The piece of paper turns red.

This argument is valid—if the premises are true, then the conclusion must be true as well. Consequently, if the conclusion is false, one or more of the premises must be false. But if the paper does not turn red, what is falsified is the conjunction of L and C, and not L itself. One many continue to affirm L, by claiming either that there was no blue litmus dye present or that the paper was not placed in an acid solution. Of course, there may be available independent means for ascertaining the truth of the statement about antecedent conditions. But observation that E is not the case does not, in itself, falsify L.

Duhem was interested primarily in more complicated cases in which a number of hypotheses are involved in the prediction that a phenomenon occurs. He emphasized that, even if the antecedent conditions are correctly stated for such cases, failure to observe the predicted phenomenon falsifies only the conjunction of hypotheses. To restore agreement with observation, the scientist is free to alter any one of the hypotheses that occur in the premises. He may decide, for instance, to retain one particular hypothesis as is, and replace or modify the other hypotheses in the set. To adopt such a strategy is to attribute to that one particular hypothesis the status of a convention for which the question of truth or falsity does not arise.

But although Duhem did indicate the way in which a hypothesis might be converted into a non-defeasible convention, he did not draw up a list of specific hypotheses which should be interpreted as nothing but conventions. He believed that, when disconfirming evidence turns up, the decision about which assumptions of a theory are to be modified should be left to the good judgement of scientists. And he indicated that a necessary condition for the exercise of good judgement is a dispassionate, objective attitude.

In some cases, there may be good reasons for making changes in one of the assumptions of a theory rather than another. This would be so, for instance, if one assumption occurs in a number of confirmed theories, whereas a second assumption occurs only in the theory under consideration. But there is nothing in the logic of disconfirmation that pinpoints the erroneous part of the theory.

Duhem applied his analysis of the logic of disconfirmation to the idea of a "crucial experiment". Francis Bacon had suggested that there do exist crucial experiments, or "Instances of the Fingerpost", which conclusively decide the issue between competing theories. In the nineteenth century, it was widely supposed that Foucault's determination that the velocity of light is greater in air than in water was a crucial experiment. The physicist Arago, for instance, claimed that Foucault's experiment demonstrated, not only that light is *not* a stream of emitted particles, but also that light *is* a wave motion.

Duhem pointed out that Arago was wrong on two counts. In the first place, the Foucault experiment falsifies only a set of hypotheses. Within the corpuscular theories of Newton and Laplace, the prediction that light moves faster in water than in air is deduced only from a group of propositions. The emission hypothesis, which likens light to a swarm of projectiles, is but one of these premises. There are, in addition, propositions about the interactions of the emitted corpuscles and the media through which they travel. Supporters of the corpuscular theory, confronted with Foucault's result, could have decided to retain the emission hypothesis and make adjustments in the other premises of the corpuscular theory. And in the second place, even if every assumption of the corpuscular theory except the emission hypothesis were known to be true on other grounds, the Foucault experiment still would not prove that light is a wave motion. Neither Arago nor any other scientist could demonstrate that light must be either a stream of emitted corpuscles or a wave motion. There may be a third alternative. Duhem emphasized that an experiment would be "crucial" only if it conclusively eliminated every possible set of explanatory premises save one. He was correct to insist that there can be no such experiments.[8]

Poincaré's Conventionalism

It was Henri Poincaré who spelled out most forcefully the implications of a conventionalist view of the general principles of science. Poincaré dissociated Whewell's claim that certain scientific laws come to be *a priori* truths, from the Kantian epistemology to which Whewell appealed to justify the *a priori* status of these laws. For Poincaré, there is no question of the existence of a set of immutable Ideas which somehow invest scientific laws with necessity. Poincaré maintained that the fact that a scientific law is held to be true independently of any appeal to experience merely reflects the implicit decision of scientists to use the law as a convention that specifies the meaning of a scientific concept. If a law is *a priori* true, it is because it has been stated in such a way that no empirical evidence can count against it.

Two Uses of the Laws of Mechanics

The law of inertia, for example, is not subject to straightforward confirmation or refutation by empirical evidence. In Poincaré's formulation, the "generalized inertial principle" specifies that the acceleration of a body depends only on its position, and on the positions and velocities of neighbouring bodies.[9] Poincaré observed that a *decisive* test of this principle would require that, after a certain period of time, each body in the universe reassume the position and velocity it had had at some particular earlier time. But such a test cannot be made. The most that can be accomplished is to examine the behaviour of groups of bodies which are "reasonably isolated" from the remainder of the universe. Needless to say, failure to observe the predicted motions within a supposedly isolated system would not falsify the generalized inertial principle. Discrepancies could be attributed to incomplete isolation of the system. The calculations could be repeated, taking into account the positions and velocities of additional bodies. There is no limit to the number of revisions of this kind that could be made.

Poincaré concluded that the generalized inertial principle may be taken to be a convention which stipulates the meaning of the phrase 'inertial motion'. On this view, 'inertial motion' *means* 'motion of a body such that its acceleration depends only on its position and the positions and velocities of neighbouring bodies'. By definition, any body whose motion is not calculated correctly from data on its position and the positions and velocities of a set of neighbouring bodies, is not a body in inertial motion.

However, although Poincaré held that the generalized inertial principle can be, and is, used as a convention which implicitly defines the phrase "inertial motion", he also held that the principle can be used as an empirically significant generalization which holds approximately for 'almost isolated' systems. Poincaré made a similar analysis of the cognitive status of Newton's other two laws of motion. On the one hand, these laws function as conventional definitions of 'force' and 'mass'. On the other hand, given procedures for measuring space, time, and force, the laws are generalizations approximately confirmed for "almost isolated" systems.

Thus is would be incorrect to attribute to Poincaré the view that general scientific laws are *nothing but* conventions which define fundamental scientific concepts. These laws do have a legitimate function as conventions, but they also have a legitimate function as empirical generalizations. Commenting on the laws of mechanics, Poincaré declared that they

present themselves to us under two different aspects. On the one hand, they are truths founded on experiment and approximately verified so far as concerns almost isolated systems. On the other hand, they are postulates applicable to the totality of the universe and regarded as rigorously true.[10]

Poincaré noted that, in the course of development of science, certain laws come to display these two aspects. Initially these laws are employed solely as experimental generalizations. For instance, a law might state a relation between terms *A* and *B*. Taking note that the relation holds only approximately, scientists may introduce term *C* which, by definition, has the relation to *A* which is expressed by the law. The original experimental law now has been subdivided into two parts: an *a priori* principle that states a relation between *A* and *C*, and an experimental law that states a relation between *B* and *C*.[11]

When implicitly defined by Newton's laws of motion, the terms 'inertial motion', 'force', and 'mass' are terms of the same type as *C*. Poincaré held that it is a matter of convention that these terms are taken to be defined by Newton's laws. No empirical evidence could prove that the stated relation of terms *A* and *C* is false. But this is not to say that the choice of definition is arbitrary. Poincaré insisted that the introduction of conventions into physical theory is justified only if it proves fruitful in subsequent research.[12]

The Choice of a Geometry to Describe "Physical Space"

Poincaré also maintained that it is a matter of convention which pure geometry is employed to describe spatial relations among bodies. However, he predicted that scientists will continue to select Euclidean geometry because it is the simplest to apply.

In the nineteenth-century, the mathematician Carl Gauss performed an experiment to confirm the Euclidean description of spatial relations. He measured the angular sum of a triangle formed by light rays emitted from distant mountain peaks. Gauss found that, within the limits of accuracy of his surveying equipment, there was no deviation from the Euclidean value of 180 degrees.

But even if Gauss had found an appreciable deviation from 180 degrees, this would not have proved that Euclidean geometry is inapplicable to spatial relations on the surface of the earth. Any deviation from the Euclidean value could be attributed to a "bending" of the light rays used to make the sightings.

Poincaré called attention to the fact that the application of a pure geometry to experience necessarily involves hypotheses about physical phenomena, such as the propagation of light rays, the properties of measuring rods, and the like. Poincaré emphasized that the application of a pure geometry to experience, like every physical theory, has an abstract component and an empirical component. When a physical geometry is not in agreement with observations, agreement may be restored either by substituting a different pure geometry—a different axiom system—or by modifying the associated physical hypotheses. Poincaré believed that, confronted with such a choice, scientists invariably would choose to modify the physical hypotheses and to retain the more convenient Euclidean pure geometry.[13]

But as Hempel has pointed out, in certain cases greater overall simplicity may be achieved by adopting a non-Euclidean geometry and retaining unchanged the associated physical hypotheses. According to Hempel, Poincaré was mistaken to restrict considerations of complexity to pure geometries alone. What counts is the complexity of the conjunction of a pure geometry and the associated physical hypotheses.[14]

Popper on Falsifiability as a Criterion of Empirical Method

Karl Popper resolved to take seriously the conventionalist point of view. He noted that it always is possible to achieve agreement between a theory and observational evidence. If certain evidence is inconsistent with consequences of the theory, a number of strategies may be pursued to "save" the theory. The evidence may be rejected outright, or it may be accounted for either by adding auxiliary hypotheses or by modifying the rules of correspondence.[15]* These

---

* Rules of correspondence are semantical rules, or "dictionary entries" (Campbell), which link the axioms of a theory to statements of empirically determined magnitudes.

strategies may introduce a staggering degree of complexity into a theoretical system. Nevertheless, evasion of falsifying evidence in these ways always is possible.

According to Popper, proper empirical method is continually to expose a theory to the possibility of being falsified. He concluded that the way to combat conventionalism is to make a decision not to employ its methods. Consistent with this conclusion, he proposed a set of methodological rules for the empirical sciences. The supreme rule is a criterion of adequacy for all other rules, much as Kant's categorical imperative is a criterion of adequacy for moral norms. This supreme rule states that all rules of empirical method

must be designed in such a way that they do not protect any statement in science against falsification.[16]

On the question of adding auxiliary hypotheses to a theory, for instance, Popper suggested that only those hypotheses be admitted which increase the degree of falsifiability of the theory. He contrasted, in this respect, Pauli's exclusion principle, and the Lorentz contraction hypothesis.[17] Pauli's principle was an addition to the Bohr–Sommerfeld theory of the atom. Pauli postulated that no two electrons in a given atom can have the same set of quantum numbers. For example, two electrons in an atom may differ in orbital angular momentum or in spin direction. Addition of this exclusion principle to the then current theory of atomic structure enabled many additional predictions to be made about atomic spectra and chemical combination. The Lorentz contraction hypothesis, on the other hand, did not increase the degree of falsifiability of the ether theory to which it was appended. Lorentz suggested that all bodies on the earth undergo a minute contraction in the direction of the earth's motion through the surrounding ether. By means of this hypothesis, he was able to account for the result of the Michelson–Morley experiment. Michelson and Morley had shown that the round-trip velocity of light is the same in all directions on the earth's surface. This experimental result was inconsistent with the ether theory, according to which the round-trip velocity should be lower in the direction of the earth's motion through the ether, than in a direction perpendicular to this motion. The Lorentz contraction hypothesis restored agreement between ether theory and experiment, but it did so in an *ad hoc* manner. No further predictions were drawn from the augmented ether theory. Popper cited the Lorentz hypothesis as an auxiliary hypothesis which should be excluded from empirical science by the falsifiability criterion.

A hypothesis that is exposed to the possibility of falsification satisfies Popper's demarcation criterion. It has qualified to be included in the realm of permissible scientific discourse. To be acceptable, a hypothesis must satisfy a further requirement. It must withstand tests designed to refute it.

Popper distinguished tests from mere instances. A test is a serious attempt at refutation. It involves a comparison between a deductive consequence of a hypothesis and a "basic statement" that records an observation.* A "basic statement" describes the occurrence of an intersubjectively observable event within a specified region of space and time.

Popper conceded that basic statements are not incorrigible. We may be mistaken about the occurrence of events. Nevertheless it is necessary to take some basic statement to be true if a hypothesis is to be put to the test. Thus there is an element of conventionalism in the testing of hypotheses. Popper declared that

the empirical basis of objective science has thus nothing 'absolute' about it. Science does not rest upon rock-bottom. The bold structure of its theories rises, as it were, above a swamp. It is like a building erected on piles. The piles are driven down from above into the swamp, but not down to any natural or 'given' base; and when we cease our attempts to drive our piles into a deeper layer, it is not because we have reached firm ground. We simply stop when we are satisfied that they are firm enough to carry the structure, at least for the time being.[18]

Popper suggested that the acceptability of a law or theory is determined by the number, diversity, and severity of tests it has passed. This is persuasive as a qualitative account. Most philosophers of science agree that a test of the law of refraction that includes various angles of incidence and numerous pairs of media is more adequate than a test restricted to the air—water interface at 30 degrees. There is general agreement as well that the discovery at the eclipse expedition of 1919 that light from distant stars is bent by the sun was a severe test of the General Theory of Relativity.[19]

---

* More precisely, it is the deductive consequence of the *conjunction* of hypothesis, statements about relevant conditions, and perhaps auxiliary hypotheses that is compared to an observation report.

It is easy to cite examples of severe tests. However, it is difficult to measure test-severity. Popper acknowledged this. He noted that severity depends on the ingenuity of an experimental arrangement, the accuracy and precision of the results achieved, and the extensiveness of connections binding the hypothesis under test to other theoretical assumptions.

Nevertheless, Popper sought to develop a quantitative measure of acceptability by reference to a concept of verisimilitude (approximation to truth). He held that the statements derivable from a theory may be divided into those that are true (its "truth-content") and those that are false (its "falsity-content"). On the assumption that the truth-contents and falsity-contents of theories T_1 and T_2 are comparable, Popper put forward the following definition of 'comparative verisimilitude':

T_2 is more closely similar to the truth, or corresponds better to the facts, than T_1, if and only if either (a) the truth-content but not the falsity-content of T_2 exceeds that of T_1, or (b) the falsity-content of T_1, but not its truth-content, exceeds that of T_2.[20]

Popper's definition is inappropriate. Tichý[21] and Miller[22] proved that if T_1 and T_2 both are false, then neither condition (a) nor condition (b) can be fulfilled. But the point of introducing verisimilitude is to allow one to say that one false theory (e.g. Newton's Theory of Gravitational Attraction) is "closer to the truth" than a second false theory (e.g. Galileo's Theory of Free Fall). Popper conceded that his initial definition of 'comparative verisimilitude' is inadequate. Unfortunately, subsequent attempts by Popper and others to amend the definition have not been successful.[23]

Popper viewed the history of science as a sequence of conjectures, refutations, revised conjectures, and additional refutations. Proper scientific procedure is to expose conjectures to the most severe tests that can be devised. If a conjecture passes a test, then it has received "corroboration". Popper insisted that corroboration is a "backward-looking" appraisal. The achievement of corroboration does not justify a belief that a hypothesis is true, or approximately true. Popper consistently has opposed the appeal to inductive arguments to justify hypotheses. On his view, it is incorrect to argue that because hypothesis H passed tests $t_1 \ldots t_n$, it is probable that H will pass test t_{n+1}.

However, Popper also has appealed frequently to an analogy

drawn from the theory of organic evolution. A well-corroborated theory has demonstrated its "fitness to survive". This evolutionary analogy creates a tension within Popper's anti-inductivist philosophy of science. It is important for a theory to pass tests. This is what establishes its evolutionary fitness within the history of science. But the passing of tests confers no epistemological benefit. One is not permitted to argue inductively that the passing of tests justifies a belief in the approximate truth of a theory. But then it is unclear why one should select for further applications a well-corroborated theory rather than a refuted theory. If inductive inference is disallowed then the following two directives are on a par:

1. apply T_2 because a previously successful theory is more likely than not to be successful in the future,
2. apply T_1 because a previously unsuccessful theory may stage a comeback.

Popper became aware of the difficulty. His response was to accept a "whiff of inductivism", based on the assumption that

reality, though unknown, is in some respects similar to what science tells us.[24]

Given this realist assumption,

we can argue that it would be a highly improbable coincidence if a theory like Einstein's could correctly predict very precise measurements not predicted by its predecessors unless there is 'some truth' in it.[25]

Critics of Popper have maintained that to accept this "whiff of inductivism" is to abandon the anti-inductivist standpoint altogether.[26]

Notes

[1] George Berkeley, 'Of Motion', in *The Works of George Berkeley*, ed. A. A. Luce and T. E. Jessop (London: Thomas Nelson, 1951), iv. 50.

[2] Ibid. 48–9.

[3] Ernst Mach. *The Science of Mechanics*, trans. T. J. McCormack (La Salle, Ill.: Open Court, 1960), 577.

[4] Ibid. 582.

[5] Ibid. 586.

[6] Mach, *History and Root of the Principle of the Conservation of Energy*, trans. P. E. B. Jourdain (Chicago: Open Court, 1911), 49.

[7] Mach, *The Science of Mechanics*, 303–4.

[8] Pierre Duhem, *The Aim and Structure of Physical Theory*, trans. Philip P. Wiener (New York: Atheneum, 1962), 186–90.

[9] Henri Poincaré, *Science and Hypothesis*, trans. G. B. Halsted (New York: Science Press, 1905), 69.

[10] Ibid. 98.

[11] Ibid. 100.

[12] Poincaré, *The Value of Science*, trans. G. B. Halsted (New York: Science Press, 1907), 110.

[13] Poincaré, *Science and Hypothesis*, 39.

[14] Carl Hempel, 'Geometry and Empirical Science', *American Mathematical Monthly*, 52 (1945), 7–17; repr. in H. Feigl and W. Sellars (eds.), *Readings in Philosophical Analysis*, 238–49.

[15] Karl Popper, *The Logic of Scientific Discovery* (New York: Basic Books, 1959), 81.

[16] Ibid. 54.

[17] Ibid. 83.

[18] Ibid. 111.

[19] See e.g. Sir Arthur Eddington, *Space, Time and Gravitation* (New York: Harper & Row, 1959), ch. 7.

[20] Popper, *Conjectures and Refutations* (New York: Basic Books, 1963), 233.

[21] Pavel Tichỳ, 'On Popper's Definition of Verisimilitude', *Brit. J. Phil. Sci.* 25 (1974), 155–60.

[22] David Miller, 'Popper's Qualitative Theory of Verisimilitude', *Brit. J. Phil. Sci.* 25 (1974), 178–88.

[23] See e.g. Anthony O'Hear, *Karl Popper* (London: Routledge & Kegan Paul, 1980), ch. 3.

[24] Popper, 'Replies to My Critics', in *The Philosophy of Karl Popper*, ed. P. A. Schilpp (La Salle, Ill.: Open Court, 1974), ii. 1192.

[25] Ibid. 1192.

[26] See e.g. W. H. Newton-Smith, *The Rationality of Science* (London: Routledge & Kegan Paul, 1981), ch. 3; O'Hear, *Karl Popper*, ch. 4; Wesley Salmon, 'Rational Prediction', *Brit. J. Phil. Sci.* 32 (1981), 115–25.

12

Logical Reconstructionist Philosophy of Science

PERCY WILLIAMS BRIDGMAN (1882–1961) was a physicist, a Nobel prize winner, who conducted pioneering investigations of the properties of matter under high pressures. His experimental determinations included the electrical and thermal properties of various substances at pressures as high as 100,000 atmospheres. In 1939 he closed his high-pressure laboratory at Harvard to visitors from totalitarian countries, an act that produced controversy within the academic community. Bridgman championed a methodological orientation known as operationalism, in which emphasis is placed on operations performed to assign values to scientific concepts.

CARL HEMPEL (1905–   ) is a German-born philosopher who studied at Göttingen, Heidelberg, and Berlin. Hempel was a member of the Berlin group that supported the aims and viewpoint of the Vienna Circle in the early 1930s. He went to the United States in 1937 and has taught at Yale and Princeton. Hempel has written important essays on the logic of scientific explanation and the structure of theories, a number of which essays are included in *Aspects of Scientific Explanation* (1965).

ERNEST NAGEL (1901–87) was born in Czechoslovakia, went to the United States in 1911, and has spent nearly all his academic career as Professor of Philosophy at Columbia. Nagel was one of the first American philosophers to take sympathetic account of the work of the Vienna Circle. His book *The Structure of Science* (1960) contains incisive analyses of the logic of scientific explanation, nomic universality, causality, and the structure and cognitive status of theories.

A Hierarchy of Language Levels

After the Second World War, philosophy of science emerged as a distinct academic discipline, complete with graduate programmes and a periodical literature. This professionalization occurred, in part, because philosophers of science believed that there were achievements to be won and that science would benefit from them.

Post-war philosophy of science was an attempt to implement a programme suggested by Norman Campbell. In *Foundations of Science* (1919),[1] Campbell noted that recent studies of the foundations of mathematics by Hilbert, Peano, and others had clarified the nature of axiomatic systems. This development was of some importance to the practice of mathematics. Campbell suggested that a study of the "foundations" of empirical science would be of similar value to the practice of science. The "foundations" Campbell discussed include the nature of measurement and the structure of scientific theories.*

Philosophers of science who sought to develop their discipline as an analogue of foundation studies in mathematics accepted Reichenbach's distinction between the context of scientific discovery and the context of justification.[2] They agreed that the proper domain of philosophy of science is the context of justification. In

---

* Campbell's position on the structure of theories is discussed in Ch. 9, pp. 141–8.

addition they sought to reformulate scientific laws and theories in the patterns of formal logic, so that questions about explanation and confirmation could be dealt with as problems in applied logic.

The great achievement of logical reconstructionism was a new understanding of the language of science. The language of science comprises a hierarchy of levels, with statements that record instrument readings at the base, and theories at the apex.

Logical reconstructionist philosophers of science drew several important conclusions about the nature of this hierarchy:

1. Each level is an "interpretation" of the level below;
2. The predictive power of statements increases from base to apex:
3. The principal division within the language of science is between an "observational level"—the bottom three levels of the hierarchy—and a "theoretical level"—the top level of the hierarchy. The observational level contains statements about "observables" such as 'pressure' and 'temperature'; the theoretical level contains statements about "non-observables" such as 'genes' and 'quarks';
4. Statements of the observational level provide a test-basis for statements of the theoretical level.

Language Levels in Science

Level	Content	For example
Theories	Deductive systems in which laws are theorems	Kinetic molecular theory
Laws	Invariant (or statistical) relations among scientific concepts	Boyle's Law ('$P \propto 1/V$')
Values of concepts	Statements that assign values to scientific concepts	'$P = 2.0$ atm.' '$V = 1.5$ lit.'
Primary experimental data	Statements about pointer readings, menisci, counter clicks, *et al*.	'Pointer p is on 3.5.'

Operationalism

In analyses dating from 1927, P. W. Bridgman emphasized that every bona fide scientific concept must be linked to instrumental procedures that determine its values.[3] Bridgman was impressed by Einstein's discussion of the concept of simultaneity.

Einstein had analysed the operations involved in judging that two events are simultaneous. He noted that a determination of simultaneity presupposes a transfer of information by means of some signal from the events in question to an observer. But the transfer of information from one point to another takes a finite period of time. Thus, in the case that the events in question occur on systems which are moving with respect to one another, judgements of simultaneity depend on the relative motions of the systems and the observer. Given a particular set of motions, observer Lynx on system 1 may judge that event x on system 1 and event y on system 2 are simultaneous. Observer Hawk on system 2 may judge otherwise. And there is no preferred standpoint from which to determine that Lynx is correct and Hawk incorrect, or vice versa. Einstein concluded that simultaneity is a relation between two or more events and an observer, and is not an objective relation between events.

Bridgman declared that it is the operations by which values are assigned that give empirical significance to a scientific concept. He noted that operational definitions link concepts to primary experimental data *via* the schema

$$(x) \, [Ox \supset (Cx \equiv Rx)]*$$

Given an operational definition, and the appropriate primary experimental data, one can deduce a value for the concept. Consider a case in which the presence of an electrically charged body is determined by operations with an electroscope:

$$(x) \, [Nx \supset (Ex \equiv Dx)]$$
$$Na$$
$$\underline{Da}$$
$$\therefore \quad Ea$$

where Nx = x is a case in which an object is brought into proximity to a neutral electroscope.

Ex = x is a case in which the object is electrically charged, and

---

* 'For all cases, if operations O are performed, then concept C applies if, and only if, results R occur.'

$Dx = x$ is a case in which the leaves of the electroscope diverge.

Since *Na* and *Da* are primary experimental data, this deductive argument enables the scientist to mount, as it were, from primary experimental data—the level of the "directly observed"—to the level of scientific concepts, viz.,

Language Level	For example
Statements that Assign Values to Scientific Concepts	*Ea*
Operational Schema	$(x) [Nx \supset (Ex \equiv Dx)]$
Primary Experimental Data	*Na, Da*

Bridgman insisted that if no operational definition can be specified for a concept, then the concept has no empirical significance and is to be excluded from science. Such was the fate of "absolute simultaneity", and Bridgman recommended similar exclusion for Newton's "Absolute Space" and Clifford's speculation that, as the solar system moves through space, both measuring instruments and the dimensions of objects measured contract at the same rate.[4]

But although Bridgman insisted that links be established between statements about theoretical terms and the observational language in which the results of measurement are recorded, he acknowledged that the links may be complex indeed. One of Bridgman's examples is the concept of stress within a deformed elastic body. Stress cannot be measured directly, but it can be calculated by means of a mathematical theory from measurements made on the surface of the body. Thus, for the concept stress, the operations performed include 'paper and pencil' operations. No matter. Given the formal relationship between 'stress' and 'strain', and the results of instrumental operations performed on the surface of the body, a value of stress follows deductively. This suffices to qualify stress as a permissible concept from the operationalist standpoint.

In his post-war writings, Bridgman emphasized two limitations of operational analysis.[5] One limitation is that it is not possible to specify all the circumstances present when an operation is performed. A compromise must be effected between the requirement of intersubjective repeatability and the desirability of a full elaboration of conditions under which an operation is performed.

Scientists have antecedent beliefs about which factors are relevant to the determination of the values of a quantity, and they proceed on the assumption that it is safe to ignore numerous "irrelevant" factors in the repetition of a given type of operation to measure that quantity. For example, scientists perform operations with manometers to determine the pressure of gases without taking into account the intensity of illumination in the room or the extent of sunspot activity. Bridgman observed that the exclusion from consideration of certain factors can be justified only by experience, and cautioned that an extension of operations into new areas of experience may require taking into consideration factors previously ignored.

A second limitation of operational analysis is the necessity to accept some unanalysed operations. For practical reasons, the analysis of operations in terms of more basic operations cannot proceed indefinitely. For example, the concept "heavier than" may be analysed in terms of operations with a beam balance. These operations may in turn be analysed further by specifying methods for constructing and calibrating balances. But provided that standard precautions about parallax are observed, scientists assume that determination of the position of the pointer on the balance scale is an operation that does not call for further analysis.

Operations performed to measure "local time" and "local length" are accepted as unanalysed operations in both classical physics and relativity physics. The "local time" of an event is its coincidence with the position of a hand on a clock. The "local length" of a body is the coincidence of its extremities with a properly calibrated, rigid rod in those cases in which there is no motion of the body relative to the rod.

Of course, the determination of coincidences in the above manner cannot guarantee that the instrument involved is functioning properly as a balance or a clock, or that the rod is a proper measure of length. Moreover, one may accept certain unanalysed kinds of coincidence-determination without committing oneself to the inflexible position that these kinds of coincidence-determination are unanalysable. Bridgman emphasized that although it is necessary to accept *some* operations as unanalysed, the decision to accept as unanalysed a particular set of operations is subject to review as our experience becomes more extensive. He noted that our experience to date has been such that no difficulties for physical theory have arisen from accepting the above coincidence-determinations as unanalysed. But he insisted that it always is possible to give a more detailed analysis of operations.[6] Thus, according to Bridgman, those currently accepted unanalysed coincidence-determinations provide for theoretical statements only a provisional anchor in the observational language.

The Deductive Pattern of Explanation

Operational schemata relate statements about scientific concepts to primary experimental data. At the next higher level, the orthodox programme is to specify the logical relations between scientific concepts and laws. The programme may be implemented from either end. Given a statement of the value of a scientific concept, one may seek to explain this fact by referring to some law. And given a law, one may seek confirming evidence among statements of the values of scientific concepts.

In a widely influential paper published in 1948, Carl Hempel and Paul Oppenheim addressed the problem of scientific explanation.[7] Commenting on an oarsman's observation that his oar is 'bent', Hempel and Oppenheim suggested that

the question 'Why does the phenomenon happen?' is construed as meaning 'according to what general laws, and by virtue of what antecedent conditions does the phenomenon occur?'[8]

The deductive pattern of explanation of a phenomenon takes the following form:

$$\begin{array}{ll} L_1, L_2, \ldots L_k & \text{General Laws} \\ C_1, C_2, \ldots C_r & \text{Statements of Antecedent Conditions} \\ \hline \therefore E & \text{Description of Phenomenon} \end{array}$$

In the case of the oarsman's observation, the general laws are the law of refraction and the law that water is optically more dense than air. The antecedent conditions are that the oar is straight and that it is immersed in water at a particular angle.

Hempel and Oppenheim made the important logical point that statements about a phenomenon cannot be deduced from general laws alone. It is necessary to include a premiss about the conditions under which the phenomenon occurs. Antecedent conditions include both the boundary conditions under which the laws are believed to hold and those initial conditions that are realized prior to, or at the same time as, the phenomenon to be explained. For instance, a deductive explanation of the expansion of a heated balloon might take the following form:

$$\left(\frac{V_2}{V_1} = \frac{T_2}{T_1}\right)$$ Gay-Lussac's Law

$$m, P = k$$

Mass and pressure   Boundary Conditions
are constant.

$$T_2 = 2T_1$$ "Initial" Conditions

$$\therefore V_2 = 2V_1$$

In the course of their discussion of the deductive pattern of explanation, Hempel and Oppenheim were careful to indicate that many bona fide scientific explanations do not fit the deductive pattern. This is the case for many explanations based on statistical laws.[9] An example given by Hempel in a subsequent essay is:

A high percentage of patients with streptococcus infections recover within 24 hours after being given penicillin.
Jones had a streptococcus infection and was given penicillin.

Jones recovered from streptococcus infection within 24 hours of receiving penicillin.[10]

This explanatory argument does not have deductive force. Rather, the premises provide only strong inductive support for the conclusion.*

Hempel thus acknowledged that subsumption under general laws may be achieved either deductively or inductively. He consistently maintained, however, that every acceptable scientific explanation involves deductive *or* inductive subsumption of an explanandum under general laws.

Nomic v. Accidental Generalizations

On the orthodox view, a successful scientific explanation subsumes its explanandum under general laws. But how can we be sure, in a

---

* A double line between premises and conclusion is used to indicate that the argument is an inductive argument.

particular case, that the premisses do include laws? We accept the following argument as a scientific explanation of a green flame test result:

> All barium-affected flames are green.
> This is a barium-affected flame.
> ∴ This flame is green.

But we deny explanatory power to the following argument:

> All the coins now in my pocket contain copper.
> This is a coin now in my pocket.
> ∴ This coin contains copper.

The two arguments have the same form. However, the former argument subsumes its explanandum under a bona fide law, whereas the latter argument subsumes its explanandum under a "merely accidental" generalization.

Orthodox theorists accepted Hume's position on scientific laws. R. B. Braithwaite, for instance, declared that

> I agree with the principal part of Hume's thesis—the part asserting that universals of law are objectively just universals of fact, and that in nature there is no extra element of necessary connexion.[11]

Braithwaite noted, however, that there are difficulties in a Humean analysis of law. One difficulty is that the Humean analysis blurs the distinction between lawlike universals and accidental universals.*

Suppose that two similar pendulum clocks are arranged to be 90° out-of-phase so that the ticks of the two clocks are in constant sequential conjunction. If scientific laws were *nothing but* statements of constant conjunction, then the following statement would be a law:

> 'For all x, if x is a tick of clock 1, then, x is a tick followed by a tick of clock 2.'

Now suppose that the pendulums of the two clocks were arrested. Does the "law" support the contrary-to-fact conditional 'If clock 1 were to tick, then this tick would be followed by a tick of clock 2'? Presumably not.

---

* Hume himself was uneasy about this distinction. See Ch. 9, pp. 110–12.

"Genuine scientific laws", on the other hand, do support contrary-to-fact conditionals. That 'All barium-affected flames are green' does support the claim that 'if that flame were a barium-affected flame, then it would be green.'

Moreover, a number of important scientific laws seem not to be about constant conjunctions at all since they refer to idealized situations that do not exist. The Ideal Gas Law is a law of this type. Even though there are no gases in which the molecules have zero extension and zero intermolecular force fields, if there were such a gas, then its pressure, volume, and temperature would be related as

$$\frac{PV}{T} = \text{constant.}$$

There is, then, a prima facie difference between lawlike universals and accidental universals. Lawlike universals support contrary-to-fact conditionals; accidental universals do not. But what does "support" mean in this context?

According to Braithwaite, this "support" results from the deductive relationship of the lawlike universal to higher-level generalizations. He suggested that a universal conditional *h* is lawlike if *h*

occurs in an established deductive system as a deduction from higher-level hypotheses which are supported by empirical evidence which is not direct evidence for *h* itself.[12]

The barium-flame-colour generalization is a deductive consequence of the postulates of atomic theory. And there is extensive confirming evidence for these postulates (over and above the colour of barium-affected flames). No such deductive relationship is known for the generalization about the two clocks.

Ernest Nagel likewise defended a Humean position on scientific laws. He maintained that lawlike generalizations can be distinguished from accidental generalizations without reference to modal notions like "necessity" and "possibility". Nagel listed four characteristics of lawlike universals:[13]

1. A universal does not acquire lawlike status solely in virtue of being vacuously true. If there are no Martians, then it is true to say that 'All Martians are green.' But truth acquired in this manner does not confer lawlike status on a statement.

There are vacuously true laws, of course. But their status as laws is determined by their logical relationship to other laws in a scientific theory.

2. The scope of predication of a lawlike universal is not known to be closed to further augmentation. The scope of predication of an accidental universal, by contrast, often is known to be closed. A case in point is 'All the coins now in my pocket contain copper.'

3. Lawlike universals do not restrict to specific regions of space or time the individuals which satisfy the antecedent and consequent conditions.

4. Lawlike universals often receive indirect support from evidence which directly supports other laws in the same scientific deductive system. For instance, if laws L_1, L_2, and L_3 are jointly derivable within an interpreted axiom system, then evidence which directly supports L_2 and L_3 provides indirect support for L_1. For example, since Boyle's Law, Charles's Law, and Graham's Law of Diffusion all are deductive consequences within the kinetic theory of gases, Boyle's Law is indirectly confirmed by evidence that confirms Charles's Law or Graham's Law. Accidental universals, by contrast, do not receive this kind of indirect support.

The Confirmation of Scientific Hypotheses

Hempel suggested in 1945 that there are three phases in the evaluation of a scientific hypothesis:[14]

1. Accumulating observation reports which state the results of observations or experiments;

2. Ascertaining whether these observation reports confirm, disconfirm, or are neutral toward, the hypothesis; and

3. Deciding whether to accept, reject, or suspend judgement on the hypothesis in the light of this confirming or disconfirming evidence.

Hempel outlined a programme of research for the second and third of these phases. Phase 2 is the problem of confirmation. Hempel maintained that this is a problem in applied logic. Both observation reports and hypotheses are sentences, and relations between sentences may be expressed in the categories of formal logic. What needs to be done is to formulate a definition of '*o* confirms *H*' in terms of logical concepts such as consistency and entailment. Armed with a suitable definition, the philosopher of science would then be able to decide whether a particular observation report confirms a hypothesis.

Qualitative Confirmation: The Raven Paradox

Hempel pointed out in 1945 that 'qualitative confirmation' is a paradoxical notion.[15] Consider the relationship between the hypothesis 'all ravens are black' and statements that record evidence. Our intuitions are that a black raven provides support for the hypothesis whereas an orange raven would refute the hypothesis. So far, so good. But the following propositions are all logically equivalent:

(1) $(x) (Rx \supset Bx)$
(2) $(x) (\sim Rx \vee Bx)$
(3) $(\sim Bx \supset \sim Rx)$

It seems plausible to hold that if an observation report confirms a generalization, then it also confirms every sentence logically equivalent to it. But a black shoe $(\sim Ra \cdot Ba)$ confirms (2),* and a white glove $(\sim Ra \cdot \sim Ba)$ confirms (3). If an Equivalence Condition is accepted, then the raven hypothesis is confirmed by both the black shoe and the white glove. This a paradoxical result. It suggests that it would be appropriate to practice ornithology indoors.

Hempel emphasized that the "Raven Paradox" results when four principles are affirmed. These Principles are:

1. The Principle of Instance Confirmation (Nicod's Criterion).
2. Equivalence Condition.
3. The assumption that many important scientific laws are universal conditionals properly symbolized '$(x) (Ax \supset Bx)$'.
4. Our intuitions about what should count as confirming instances.

To dissolve the paradox, it is necessary to reject one or more of the four principles.

Hempel maintained that the Principle of Instance Confirmation

---

* Since (2) States that, 'given anything in the universe, either it is not a raven or it is black', it is appropriate to take a specific black non-raven to be an "instance" of (2).

and the Equivalence Condition are deeply embedded in scientific practice, and that many important scientific laws are represented correctly as universal conditionals. His own position on the Raven Paradox was that we are misguided by our intuitions. In the first place, we judge wrongly that 'All ravens are black' is exclusively "about" ravens. But this is not the case. It is, rather, "about" all the objects in the universe. It asserts that 'given anything in the universe, if it is a raven, then it is black'. An equivalent formulation is $(x) [\sim Rx \lor Bx]$, which asserts that 'given anything in the universe, either it is not a raven, or it is black'.

A second reason why our intuitions about confirmation often are wrong is that we tacitly appeal to our background knowledge when judging whether an evidence statement confirms a generalization. For instance, we know that there are many more non-black objects than ravens. And we also know that our chance of finding a disconfirming case—$Ra \cdot \sim Ba$—is greater if we examine ravens for colour than if we examine non-black objects for "ravenhood". Since the risk of falsification is greater if we focus on the class of ravens, we regard a case in which a raven has passed the test—$Ra \cdot Ba$—as a confirming case. On the other hand, we are not impressed when a non-black object passes the test—$\sim Ba \cdot \sim Ra$.

But suppose we knew that there were just ten objects in the universe, that nine of the ten were ravens, and that only one of the ten was not black. If this were our background knowledge then our intuitions about confirmation would be different. We would seek confirming evidence for 'All ravens are black' by examining the one non-black object for ravenhood.

Hempel concluded that the relationship between generalizations and their confirming instances is not paradoxical to the properly educated intuition. If one keeps in mind the logical form of a universal generalization, and if one excludes background knowledge about relative class-sizes, then there is no paradox. Hempel insisted that statements about black ravens, statements about black shoes, and statements about white gloves all count as confirming evidence for 'All ravens are black.'[16]

Carnap On Quantitative Confirmation

Rudolf Carnap concluded that the prospects of a theory of qualitative confirmation were unpromising. He sought, instead, to formulate a theory to measure the *degree* of confirmation afforded hypothesis *H* by evidence *e*. Carnap's project was to:

1. Specify the structure and vocabulary of an artificial language within which '$c(H,e) = k$' can be defined;*
2. Enlist the resources of the mathematical theory of probability to assign values to k; and
3. Argue that the calculated values are consistent with our intuitions about confirmation.[17]

Unfortunately, the '*c*-functions' developed by Carnap assign the value '$c = 0$' to those universal conditionals for which infinitely many substitution instances are possible. This is counter-intuitive. We believe, for example, that the degree of confirmation of the law of gravitational attraction on the evidence is considerably greater than zero.

Carnap acknowledged this. But he insisted that when a scientist uses a universal generalization, she need not commit herself to the truth of the generalization over a large number of instances. It suffices that the generalization hold true in the next instance. Carnap was able to show that this "next-instance confirmation" of a universal generalization approaches 1 as sample size increases, provided that there are no refuting instances in the sample.[18] Opinions were divided on the appropriateness of this shift of emphasis from 'confirmation' to 'next-instance confirmation'.

The Structure of Scientific Theories

Post-war analyses of the structure of theories were based on Campbell's distinction between an axiom system and its application to experience.† Rudolf Carnap restated the "hypothesis-plus-dictionary" view of scientific theories in an influential essay

---

* The ingredients of the artificial language include;
  1. truth-functional connectives and quantifiers,
  2. individual constants that name individuals,
  3. primitive predicates that are finite in number, co-ordinate, and logically independent of one another, and
  4. rules of sentence formation and deductive inference.

† Campbell's view of theories is discussed above, pp. 141–8.

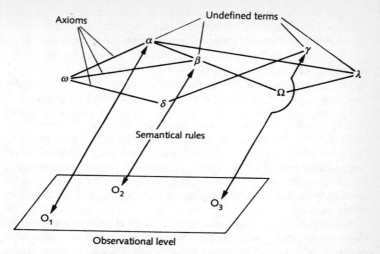

Hempel's "Safety-Net" Image of Theories

published in the *International Encyclopedia of Unified Science* in 1939. He declared that

any physical theory, and likewise the whole of physics, can ... be presented in the form of an interpreted system, consisting of a specific calculus (axiom system) and a system of semantical rules for its interpretation.[19]

This claim was repeated by Philipp Frank and Carl Hempel in subsequent essays in the same encyclopedia.[20]

Hempel's version of the hypothesis-plus-dictionary view bears some resemblance to safety-nets used for the protection of trapeze artists. The axiom system is a net supported from below by rods anchored at the observational level of scientific language.[21]

Following Campbell, Hempel observed that it is not necessary that every knot in the net have a point of support among the statements of the observational level. This being the case, the question naturally arises, under what conditions is the net securely anchored? How can it be known whether there is a sufficient number of links of adequate strength between the net and the plane of observation? The strength of the anchoring relation is greatest for "mathematical theories" in which each term of the calculus is assigned a semantical rule. Physical geometry is an example of a theory of this type. Each of the terms of the calculus— 'point', 'line', 'congruence' ... —is correlated with physical operations. At the other extreme, one could imagine a "mechanical theory" whose calculus was linked to observables by a single semantical rule. Would such a 'theory' be empirically significant?

Hempel suggested that a satisfactory answer could be given to this question if there were available an adequate theory of confirmation. According to Hempel, an adequate theory of confirmation would contain rules such that, for every theorem (T) and every sentence of the observation language reporting evidence (E), the rules confer a specific degree of confirmation on T with respect to E. A theory to which confirmation rules applied in this way would qualify as empirically significant. The semantical rules of such a theory would be of sufficient strength to anchor its calculus. However, Hempel conceded that no theory of confirmation presently available was adequate for the indicated purpose.[22] Consequently, his proposal (in 1952) to measure the adequacy of the empirical interpretation of calculi by a theory of confirmation had the status of a programme for future enquiry.

Theoretical terms for which there are no dictionary entries nonetheless are assumed to be empirically significant. R. B. Braithwaite suggested that empirical significance is conferred upwards from statements about observables to axioms.[23] In the quantum theory, for instance, it is theorems about electron charge densities, scattering distributions, and the like, that confer empirical significance upon the 'Ψ-function'. Noretta Koertge noted that the logical reconstructionist position is that empirical meaning seeps upwards *via* "capillary action" from the soil of the observational level of scientific language.[24]

Theory Replacement: Growth by Incorporation

It was the orthodox position that to explain a phenomenon is to show that its description follows logically (usually deductively) from laws and statements of antecedent conditions. Similarly, to explain a law is to show that it follows logically from other laws.[25]

Applied to the history of science, this concern with a logical reconstruction of the relation between laws was reflected in an emphasis on "growth by incorporation". Ernest Nagel observed that

the phenomenon of a relatively autonomous theory becoming absorbed by, or reduced to, some other more inclusive theory is an undeniable and recurrent feature of the history of modern science.[26]

Nagel distinguished two types of reduction. The first type is homogeneous reduction, in which a law subsequently is incorporated into a theory which utilizes "substantially the same" concepts that occur in the law. He suggested that the "absorption" of Galileo's law of falling bodies into Newtonian mechanics is a reduction of this type.[27] According to Nagel, Galileo's law has been reduced to, and is explained by, the principles of Newtonian mechanics.

A second, more interesting, type of reduction is the deductive subsumption of a law by a theory that lacks some of the concepts in which the law is expressed. Frequently, the law subsumed refers to macroscopic properties of objects and the reducing theory refers to the micro-structure of the objects. An example to which Nagel devoted some attention is the reduction of classical thermodynamics to statistical mechanics.[28] There occur in the laws of classical thermodynamics concepts which are not included among the concepts of statistical mechanics. Among these concepts are "temperature" and "entropy". Maxwell and Boltzmann, nevertheless, succeeded in deducing the laws of classical thermodynamics from premises which include statistical laws about the motions of molecules.

Reflecting on this typical case of heterogeneous reduction, Nagel sought to uncover the necessary and sufficient conditions for the reduction of one branch of science to another. He cautioned that conditions for reduction can be formulated only for branches of science that have been formalized. One requirement for formalization is that the meanings of the terms which occur in the theories in question are fixed by rules of usage appropriate to each discipline. Given that this is the case, and that the relations of logical dependence within each theory have been stated, the following are necessary conditions for the reduction of T_2 to T_1.[29]

Formal Conditions for Reduction

1. Connectability: for each term which occurs in T_2 but not in T_1, there is a connecting statement which links the term with the theoretical terms of T_1.

2. Derivability: the experimental laws of T_2 are deductive consequences of the theoretical assumptions of T_1.

Non-Formal Conditions for Reduction

3. Empirical Support: the theoretical assumptions of T_1 are supported by evidence over and above that evidence which supports T_2.

4. Fertility: the theoretical assumptions of T_1 are suggestive of further development of T_2.

Progress by Incorporation

Successful reduction is incorporation. One theory is absorbed into a second theory which has a broader scope. This suggests that progress in science is much like the creation of an expanding nest of Chinese boxes.

In essays written in the 1920s and subsequently, Niels Bohr championed this view of scientific progress. He maintained that the Chinese-box view is a fruitful methodological application of the Correspondence Postulate.*

---

* The Correspondence Postulate was an axiom of Bohr's theory of the hydrogen atom (1913). In order to account for the observed spectrum of hydrogen, Bohr suggested that the hydrogen electron can exist only in certain stable orbits, the angular momenta of which are given by $m\,v\,r = \dfrac{n\,h}{2\pi}$, where m is the mass of the electron, v is its velocity, r is the radius of its orbit, h is Planck's constant, and n is a positive integer. Transition from one stable orbit to another is accompanied by the emission or absorption of energy (e.g., the transition from $n = 3$ to $n = 2$ produces the first spectral line in the Balmer Series). The Correspondence Postulate stipulates that, in the limit as n approaches infinity and the electron no longer is bound to the nucleus, the electron obeys the laws of electrodynamics.

Encouraged by the success of his theory of the hydrogen atom, Bohr maintained that a generalized version of the Correspondence Postulate is a criterion of acceptability for quantum mechanical theories. According to Bohr, whatever the form of a theory of the quantum domain, it must be in asymptotic agreement with classical electrodynamics in the region for which the classical theory has proved adequate.[30]

To apply the Correspondence Principle as a criterion of acceptability is to require of every candidate to succeed a theory *T* that (1) the new theory has a greater testable content than *T*, and (2) the new theory is in asymptotic agreement with *T* in the region for which *T* is well confirmed.

Joseph Agassi has expressed this methodological extension of the Correspondence Postulate as follows:

there are two acknowledged methodological demands which can be made of any newly proposed theory: it should yield the theory it comes to replace as a consequence or as a first approximation and also as a special case. The first demand amounts to nothing more than the demand that the new theory explain the success which the preceding theory had. The second demand amounts to the requirement that the new theory be more general and independently testable.[31]

Notes

[1] N. R. Campbell, *Foundations of Science* (New York: Dover Publications, 1957), 1–12.

[2] Hans Reichenbach, *The Rise of Scientific Philosophy* (Berkeley, Calif.: University of California Press, 1951), 231. This distinction had been made earlier by John Herschel. Herschel's use of the distinction is discussed in Ch. 9, Sect. II of the present work.

[3] P. W. Bridgman, *The Logic of Modern Physics* (New York: The Macmillan Company, 1927); *The Nature of Physical Theory* (Princeton, NJ: Princeton University Press, 1936).

[4] Bridgman, *The Logic of Modern Physics*, 28–9.

[5] Bridgman, *Reflections of a Physicist* (New York: Philosophical Library, 1950), 1–42; *The Way Things Are* (Cambridge, Mass.: Harvard University Press, 1959), Chapter 3.

[6] Bridgman, *The Way Things Are*, 51.

[7] Carl G. Hempel and Paul Oppenheim, 'Studies in the Logic of Explanation', *Phil. Sci.* 15 (1948), 135–75; repr. in Hempel, *Aspects of Scientific Explanation* (New York: Free Press, 1965), 245–95.

[8] Ibid. 246.

[9] Ibid. 250–1.

[10] Hempel, *Aspects of Scientific Explanation*, 382.

[11] R. B. Braithwaite, *Scientific Explanation* (Cambridge: Cambridge University Press, 1953), 294.

[12] Ibid. 302.

[13] Ernest Nagel, *The Structure of Science* (New York: Harcourt, Brace, & World, 1961), 56–67.

[14] Carl Hempel, 'Studies in the Logic of Confirmation', *Mind*, 54 (1945), 1–26; 97–121. Repr. in Hempel, *Aspects of Scientific Explanation* 3–46.

[15] Ibid.

[16] Ibid. 18–20.

[17] Rudolf Carnap, *Logical Foundations of Probability* (Chicago: University of Chicago Press, 1950).

[18] Ibid. 572–3.

[19] Carnap, 'Foundations of Logic and Mathematics' (1939), in *International Encyclopedia of Unified Science*, vol. i, pt. 1, ed. O. Neurath, R. Carnap, and C. Morris (Chicago: University of Chicago Press, 1955), 202.

[20] Philipp Frank, 'Foundations of Physics,' in *International Encyclopedia of Unified Science*, vol. i, pt. 2, 429–30; Hempel, 'Fundamentals of Concept Formation in Empirical Science', in *International Encyclopedia of Unified Science*, vol. ii, no. 7, 32–9.

[21] Hempel, 'Fundamentals of Concept Formation in Empirical Science', 29–39.

[22] Ibid. 39.

[23] Braithwaite, *Scientific Explanation*, 51–2, 88–93.

[24] Noretta Koertge, 'For and Against Method', *Brit. J. Phil. Sci.* 23 (1972), 275.

[25] Nagel, *The Structure of Science*, 33–42.

[26] Ibid. 336–7.

[27] Ibid. 339.

[28] Nagel, *The Structure of Science*, 342–66; 'The Meaning of Reduction in the Natural Sciences', in *Readings in Philosophy of Science*, ed. P. Wiener (New York: Charles Scribner's Sons, 1953), 535–45.

[29] Nagel, *The Structure of Science*, 345–66.

[30] Niels Bohr, 'Atomic Theory and Mechanics' (1925), in *Atomic Theory and the Description of Nature* (Cambridge: Cambridge University Press, 1961), 35–9.

[31] Joseph Agassi, 'Between Micro and Macro', *Brit. J. Phil. Sci.* 14 (1963), 26.

13

Orthodoxy under Attack

PAUL FEYERABEND (1924–) received a Ph.D. from the University of Vienna and has taught at the University of California. He is a self-professed "anarchist" who opposes the search for rules of theory-replacement and "rational reconstructions" of scientific progress. Feyerabend's position is that "anything goes" and that the mark of creativity in science is a proliferation of theories. Consistent with this orientation, his major work is titled *Against Method* (1975).

NELSON GOODMAN (1906–) received a Ph.D. from Harvard and has taught at the University of Pennsylvania, Brandeis, and Harvard. He has made important contributions to inductive logic, epistemology, and the philosophy of art. He is the author of *The Structure of Appearance* (1951), *Fact, Fiction and Forecast* (1955), and *Languages of Art* (1968).

STEPHEN TOULMIN (1922–   ) received a D.Phil from Oxford and has taught at the University of Leeds, Michigan State, the University of Chicago, and the University of California. He has written widely on topics in the history and philosophy of science, epistemology, and ethics. In recent work he has outlined a reconstruction of scientific growth in categories borrowed from the theory of organic evolution.

HERBERT FEIGL (1902–88) participated in the activities of the Vienna Circle (1924–30), as friend and associate of Schlick and Carnap. He came to the United States in 1930 to work with P. W. Bridgman. Feigl was appointed Professor of Philosophy at the University of Minnesota in 1940 and was instrumental in the founding and continuing success of the Minnesota Center for the Philosophy of Science. Feigl wrote in support of mind–body identity, Scientific Realism, and an empiricism free from metaphysics.

The Logical Reconstructionist view of science came increasingly under attack during the late 1950s and 1960s. Critics assailed the observational level-theoretical level distinction, the covering-law model of explanation, the Safety-Net image of theories, the principle of confirmation by instances, and the Chinese-box view of scientific progress.

Is There a Theory-Independent Observational Language?

Basic to the Logical Reconstructionist philosophy of science is a claim about the theory-independence of observation reports. Orthodox theorists assumed that the truth or falsity of observation reports can be decided directly without appeal to sentences of the theoretical level. It was the orthodox position that theory-independent sentences of the observational level provide bona fide tests of theories. It also was the orthodox position that the sentences of the theoretical level acquire empirical meaning from the sentences of the observational level. Thus the theoretical level is parasitic upon the observational level.

Paul Feyerabend suggested that the dependence had been mis-construed. It is observation reports that are parasitic on theories. Feyerabend called attention to the theory-dependence of observation reports by means of the following example.[1] Take L_0 to be a language in which colours are ascribed to self-luminescent objects.

Assume that L_0 contains names a, b, c . . . and colour-predicates P_1, P_2, P_3 . . . Assume also that users of this language interpret P_i terms as designating properties possessed by the objects whether or not they are observed.

Now suppose that a scientist claims that colours recorded by an observer depend on the relative velocity of observer and source. To accept this theory is to change the interpretation of the sentences of L_0. Now 'a is P_1' no longer ascribes a property to the object named. Now it asserts a relation between object and observer, a relation that depends on their relative velocity. On this new interpretation it is not meaningful to talk about the colour properties of unobserved objects. Feyerabend concluded that

the interpretation of an observation-language is determined by the theories which we use to explain what we observe, and it changes as soon as those theories change.[2]

One consequence of Feyerabend's thesis is that the observational term—theoretical term distinction is context-dependent. Peter Achinstein provided additional support for this consequence.

Achinstein surveyed the ways in which the observable—non-observable distinction is drawn in practice. Upon occasion we accept as a case of "observing X" the observation of some Y that normally accompanies X. In this sense of "observe" a forest ranger observes a fire by attending to a cloud of black smoke. And a physicist observes the passage of an electron through a cloud chamber by attending to a curved white track. We also accept as a case of "observing X" attending to an image of X produced by a mirror or a lens. Suppose we wish to observe a slice of muscle tissue. We might examine the tissue successively with the naked eye, under a microscope, under a microscope after staining and fixing, and under an electron microscope. Do we "observe" the tissue in each instance? Or is there a point in this sequence at which we have ceased to observe the tissue? Achinstein emphasized that our classification into "observables" and "non-observables" depends on the purpose of the classification.[3]

The contrast "observable—non-observable" is a context-dependent contrast. The appropriate response to the question "Is X an observable?" is to ask the questioner to specify the kind of contrast he has in mind. Given that 'X' is used in certain contexts, which other terms—'A', 'B', 'C' . . .—does the questioner take to be "non-observables"? Given this information, a comparison can be made. Consider the term 'virus-stained-and-viewed-under-an-electron-microscope' (*t*). One might classify this term as "non-observable" relative to the term 'diamond-viewed-under-an-electron-microscope', since what is "observed" in the former case is not the virus itself but the heavy molecules attached to it in the staining process. But one might classify '*t*' as observable relative to the term 'virus-stained-and-viewed-by-X-ray-diffraction', since the electron-microscope image is a likeness of the virus in a way in which the X-ray diffraction pattern is not.[4]

Additional difficulties for the observational term–theoretical term distinction were raised by Willard van Orman Quine. Quine reaffirmed and developed a thesis which had been suggested by Pierre Duhem.[5] Quine's version of Duhem's thesis is that "our statements about the external world face the tribunal of sense experience not individually but only as a corporate body".[6] Quine called attention to the following consequences of the Duhem thesis:

1. it is misleading to speak of the "empirical content" of an individual statement;
2. any statement can be retained as true provided that sufficiently drastic adjustments are made elsewhere in the system; and
3. there is no sharp boundary between synthetic statements whose truth (or falsity) is contingent upon empirical evidence, and analytic statements whose truth (or falsity) is independent of empirical evidence.[7]

If the Duhem–Quine thesis is correct, then the orthodox view of scientific theories is untenable. According to the "Safety-Net" image, for instance, the axiom system and rules of correspondence can be reformulated in diverse ways provided that the net thus created is supported by rods extending from the observational level of scientific language. In the "Safety-Net" interpretation, it is observation reports that support the rods. The orthodox position was that the truth-status of an observation report is independent of the truth-status of the statements of the interpreted axiom system. To pursue the metaphor, the points of support are there first, and the theoretician's task is to ensure that the rods are placed directly upon them.

But if Feyerabend and Quine are correct, the points of support for a theory are created by the theory itself. Observation reports have no status apart from the theoretical context in which they occur.

Doubts About the Covering-Law Model of Explanation

A corner-stone of post-war orthodoxy was that scientific explanation is a subsumption of the explanandum under general laws. On the covering-law model the explanation of individual events is an instantiation of either the DN pattern (deductive–nomological) or the IS pattern (inductive–statistical). Certain critics of the covering-law model accused Hempel of maintaining that subsumption under general laws is a sufficient condition for scientific explanation.* But Hempel did not defend this position. Indeed, he called attention to the following example suggested by S. Bromberger:

Laws	Theorems of physical geometry
Antecedent Conditions	Flagpole F stands vertically on level ground and subtends an angle of 45 degrees when viewed from ground level at a distance of 80 feet.
∴ Phenomenon	Flagpole F is 80 feet high.

Hempel conceded that the premises of this argument do not explain why the flagpole is 80 feet high.[11]

Hempel noted, moreover, that scientists often make use of "indicator laws" for purposes of prediction. He pointed out that subsumption under an indicator law may fail to explain a phenomenon. An example is

All patients with Koplik spots on the mucous lining of
the cheeks subsequently develop measles.
Jones had Koplik spots on his cheeks last week.

∴ Jones has measles today.[12]

This argument instantiates the DN pattern. However, it does not count as an explanation of Jones's measles to claim that he has

---

* Among the critics were William Dray,[8] Michael Scriven,[9] and Richard Zaffron.[10]

measles *because* he previously had spots on his cheeks. Nor does it count as an explanation of today's rainstorm to claim that it rained *because* a barometer reading decreased yesterday. "Indicator laws" are valuable for the purpose of prediction, but are not of value as premises in explanatory arguments.

Instantiation of the IS pattern is not a sufficient condition of scientific explanation either. Wesley Salmon pointed out that many arguments similar to Hempel's "streptococcus-penicillin" argument fail to explain. For example

> A high percentage of individuals with colds recover within a week after administration of vitamin C.
> Jones had a cold and took vitamin C.
>
> ∴ Jones recovered from his cold within one week after taking vitamin C.[13]

This argument is non-explanatory despite the fact that it invokes a highly probable correlation. What counts, for explanatory purposes, is whether recovery after administration of vitamin C is more likely than spontaneous recovery. Salmon insisted that what is important in statistical explanations is not high probability, but rather the "statistical relevance" of the explanatory premises.

Thus one may instantiate either of the covering-law patterns without achieving explanation. However, it still may be the case that instantiation of one of the patterns is a *necessary* condition of scientific explanation.

The status of the DN pattern was the subject of an extended debate between Hempel and Michael Scriven.[14] Scriven maintained that DN subsumption is not a necessary condition for deductive explanation. He noted that deductive explanations of events often have the form '*q* because *p*'. An example given by Scriven is 'The bridge collapsed because a bomb exploded nearby.' Scriven conceded that if this explanation is challenged, then the appropriate defence is to cite laws that correlate explosive force, distance, and the tensile properties of materials. But the relevant laws need not be stated explicitly as premises of the explanation.

Hempel replied that to select a particular set of antecedent conditions as the cause of a particular effect is to presuppose the applicability of covering laws. He maintained that to assert '*q* because *p*' is to claim that antecedent conditions of the type described by '*p*' regularly yield effects of the type described by '*q*'. It is this putative regularity that elevates '*q* because *p*' from mere sequential narrative to causal account. Hempel declared that '*q* because *p*' counts as an explanation only if there are covering laws, which conjoined with '*p*' (and perhaps other tacitly assumed antecedent conditions) imply '*q*'.[15] Hempel thus presented a strong defence of the position that DN subsumption is a necessary condition of deductive explanations.

The IS pattern proved more vulnerable to criticism. Wesley Salmon complained that the IS pattern cannot account for the occurrence of improbable events. Consider the correlation between exposure to radiation and the subsequent development of leukemia. Salmon emphasized that there is a causal relation in such cases, even though only 1 per cent of persons exposed to a certain level of radiation develop leukemia. It is the statistical relevance of the contrast "exposure *v.* non-exposure" that has explanatory force.[16]

Suppose Smith has been exposed to a low level of radiation and has developed leukemia. No IS explanation of this event is available, since the IS pattern is applicable only to highly probable correlations, and the radiation–leukemia correlation is not probable. Nor is a DN argument available to explain Smith's illness.* And yet it seems clear that one does explain Smith's illness by citing his prior exposure to radiation.

A Non-Statement View of Theories

On the orthodox view, a theory is a collection of sentences. A number of critics opposed this view. Frederick Suppe, for instance, proposed a "non-statement view" of theories.[17] On the "non-statement view" a "theory" is rather like a proposition. Consider the sentences

(1) John loves Mary.
(2) Mary is loved by John.

---

* A DN explanation is available for the (admittedly low) *probability* that Smith contract leukemia. But this explanation is not an explanation of the event in question.

Some logicians would maintain that, although the two sentences are different, they express a single proposition.* A similar relationship may be suggested between alternative formulations of quantum theory and quantum theory itself. Von Neumann had shown that Schrödinger's wave mechanics and Heisenberg's matrix mechanics are equivalent.[18] It would seem that quantum theory is "expressed" by each of these formulations much as the "proposition" or "meaning" of the John–Mary relation is "expressed" by each of the two sentences above.

Suppe suggested that a generalization of von Neumann's result provides a fruitful reinterpretation of the nature of scientific theories. On this reinterpretation, a theory is a non-linguistic entity which is related to, but different from, a set of linguistic formulations. A theory has an "intended scope", a class of phenomena to be explained. But the theory does not describe phenomena directly. Rather, it specifies a replica, an idealized physical system. The states of this idealized system are determined by values of parameters of the theory. Formulations of the theory make contrary-to-fact claims of the form 'if the phenomena were fully characterized by the parameters of the theory, then . . .'.

What then do theories explain? The Logical Reconstructionist position was that theories explain experimental laws. They do so by means of deductive arguments in which the laws are conclusions. For example, Boyle's Law may be explained by formulating a deductive argument whose premises include the axioms and rules of correspondence of the kinetic theory of gases. Orthodox theorists thus echoed Pierre Duhem's dictum that a theory explains laws by incorporating them into a deductive system. Duhem had insisted that a theory explains because it implies laws, and not because it depicts some "reality" that underlies phenomena.[19]

Wilfred Sellars complained that it is a mistake to identify explanation and implication in this way. Sellars maintained that what a theory explains is why phenomena obey particular experimental laws to the extent that they do. For example, the kinetic theory explains why a gas under moderate pressure obeys the law $\frac{PV}{T} = k$. A gas under moderate pressure behaves as if it

---

* For a discussion of the sentence–proposition distinction, see S. Gorovitz and R. G. Williams, *Philosophical Analysis* (New York: Random House, 1963), ch. 4.

were an "ideal gas", the parameters of which are specified by the theory. Sellars declared that

roughly, it is because a gas "is"—in some sense of "is"—a cloud of molecules which are behaving in theoretically defined ways...that it obeys the Boyle–Charles law.[20]

Sellars noted that the kinetic theory also explains why the behaviour of a gas diverges from $\frac{PV}{T} = k$ at high pressures. An "ideal gas" is a collection of point-masses devoid of inter-particle forces. No actual gas can be so composed. And the "idealized replica" becomes an increasingly inappropriate approximation as the pressure of a gas increases.

Goodman's "New Riddle of Induction"

In an important study published in 1953, Nelson Goodman pointed out an important difficulty for confirmation theory.[21] This difficulty is that not every generalization is supported by its positive instances. Nicod's Criterion is inadequate. Goodman noted that whether a generalization is supported by its instances depends on the nature of the property terms that occur in the generalization. He compared the following two generalizations:

(1) All emeralds are green.
(2) All emeralds are grue.

where "x is grue" if, and only if, "either x is examined before time t and is green, or t is not examined before time t and is blue".[22]

Instances of emeralds examined before t and found to be green presumably would support (2) as well as (1). But this is disturbing. Suppose t is some time today. Which generalization should we use to predict the colour of emeralds that may be discovered tomorrow? If we rely exclusively on the number of positive instances which have been in accord with the generalization prior to t, then we have no basis for preferring (1) to (2).

We believe that (1) is a lawlike generalization and that (2) is not. Goodman suggested that (2) is an "accidental" generalization of the same sort as

(3) All men now in this room are third sons.

According to Goodman, evidence that a man now in this room is a third son does not support the claim that another man now in the room also is a third son. The situation is different in the case of "genuine" or "lawlike" generalizations. For instance, evidence that an ice cube floats on water does support the claim that another ice cube will also float. Goodman maintained that the generalization about the "grueness" of emeralds resembles the "accidental" generalization about third sons with respect to its relationship to its instances. He called attention to the task of specifying criteria to distinguish those generalizations that are supported by their positive instances from those that are not.

One approach might be to subdivide predicates into those that involve spatial or temporal reference and those that do not. Then lawlike generalizations could be restricted to generalizations whose non-logical terms lack spatial and temporal reference. Presumably this would rule out the generalizations about grue emeralds and men now present in this room.

Goodman rejected this approach. He pointed out that the riddle about emeralds can be restated without using predicates that have temporal reference.[23] On the assumption that there is a finite set of individuals n, which have been examined and have been found to be green emeralds, the predicate "grue" may be defined with respect to this set of individuals:

"x is grue" if, and only if,
"either x is identical with $(a \lor b \lor c \lor \ldots n)$ and is green, or x is not identical with $(a \lor b \lor c \lor \ldots n)$ and is blue."

On this definition of "grue", it still is true that each individual which is a positive instance of generalization '(1)' also is a positive instance of generalization '(2)'.*

Goodman maintained that the way to overcome the difficulties associated with predicates like 'grue' and 'men now in this room' is to take a pragmatic–historical approach. One should begin with the record of past usage of predicates, and use this "track record"

---

* A further difficulty for this approach is that some generalizations which scientists call "laws" do involve terms that have spatial or temporal reference. An example is Kepler's First Law, which refers the elliptic orbits of the planets to the position of the sun.

to classify them. Certain predicates have participated in generalizations that have been projected successfully to account for new instances. Goodman labelled such terms "entrenched predicates".[24] 'Green', for example, is an entrenched predicate. This is because generalizations such as 'All emeralds are green' and 'All barium compounds burn with a green flame' have been projected on to additional instances. 'Grue', by contrast, is not an entrenched predicate. It has not participated in successfully projected generalizations. Of course, it might have been so used, but what counts is actual usage, and the biographies of 'grue' and 'green' are quite different.

If Goodman is correct, then lawlike status is a matter of projectibility, projectibility is a function of the comparative entrenchment of predicates, and entrenchment itself is determined by past usage. One effect of Goodman's discussion of the "New Riddle of Induction" was to "downgrade" a philosophical problem into a historical problem. To be sure, it remains for the philosopher of science to specify the criteria of projectibility. But since the criteria deal with the entrenchment of predicates, and entrenchment is determined by examining the biographies of predicates, the really important task is that performed by the historian of science.

A second effect of Goodman's discussion was to undermine the orthodox assumption that confirmation is an exclusively logical relation between sentences. In a Postscript (1964) to his 1945 essay, Hempel conceded that

the search for purely syntactical criteria of qualitative or quantitative confirmation presupposes that the hypotheses in question are formulated in terms that permit projection; and such terms cannot be singled out by syntactical means alone.[25]

Doubts about the Chinese-Box View of Scientific Progress

Feyerabend's Incommensurability Thesis

Feyerabend claimed that the traditional examples of "reduction" discussed by orthodox theorists fail to satisfy their own requirements for reduction. One such example is the supposed reduction of Galilean physics to Newtonian physics. Feyerabend noted that Nagel's condition of derivability is not fulfilled in this case. A basic law of Galilean physics is that the vertical acceleration of falling bodies is constant over any finite vertical interval near the earth's surface. But this law cannot be deduced from the laws of Newtonian physics. In Newtonian physics the gravitational attractive force, and hence the mutual acceleration, of two bodies increases with decreasing distance. The Galilean law could be derived from Newtonian laws only if the ratio distance of fall : radius of earth were o. But in cases of free fall, this ratio never is equal to zero. The Galilean relation does not follow logically from the laws of Newtonian mechanics.[26]

A second example is the supposed "reduction" of Newtonian mechanics to General Relativity Theory. Feyerabend conceded that under certain limiting conditions the equations of Relativity Theory yield values that approach those calculated within Newtonian mechanics. But this does not suffice to establish the reduction of Newtonian mechanics to General Relativity Theory. The condition of connectability is not fulfilled in this case. Consider the concept "length": in Newtonian mechanics, length is a relation that is independent of signal velocity, gravitational fields, and the motion of the observer. In Relativity Theory, length is a relation whose value is dependent on signal velocity, gravitational fields, and the motion of the observer. The transition from Newtonian mechanics to Relativity Theory involves a change of meaning of spatiotemporal concepts. 'Classical length' and 'relativistic length' are incommensurable notions,[27] and Newtonian mechanics is not reducible to General Relativity Theory. Feyerabend also maintained that classical mechanics cannot be reduced to quantum mechanics,[28] and that classical thermodynamics cannot be reduced to statistical mechanics.[29]

Hilary Putnam suggested that Nagel's Theory of Reduction can be protected against Feyerabend's criticism by means of a minor modification. We need only specify that it is a suitable approximation of the old theory that is deducible from the new one.[30]

Feyerabend replied that the original interest in reduction had been an interest in a relationship between various actual scientific theories.[31] He noted that Putnam had salvaged the Theory of Reduction only by making it inapplicable to actual cases of theory replacement.

Feyerabend claimed to have shown that the examples of reduction cited by orthodox theorists do not satisfy their own conditions for reduction. Rather, high-level theory-replacement involves changes in the meanings of those descriptive terms that occur in both theories. The successor theory reinterprets the descriptive vocabulary that previously had been in use. But observation reports that are theory-dependent in this way cannot serve as an objective basis for the evaluation of competing theories. Feyerabend concluded that high level theories are observationally incommensurable.[32]

Growth by Incorporation or Revolutionary Overthrow?

William Whewell had compared the growth of a science to the confluence of tributaries to form a river.[*] The tributary–river image is consistent with the Chinese-box view of progress-by-incorporation and the attendant philosophical interest in the problem of reduction. The tributary–river image is also consistent with Bohr's use of the Correspondence Principle as a methodological guide to theory-formation.[†]

Post-war critics of this overview complained that the tributary–river image superimposes a false continuity on the history of science. Science does not develop smoothly. Theories do not flow into one another. Rather, competition is the rule, and the replacement of one theory by another is often by revolutionary overthrow.

Stephen Toulmin pointed out that drastic conceptual changes often accompany the replacement of one inclusive theory by another.[33] Most important in the history of science have been changes in "Ideals of Natural Order". Ideals of Natural Order are standards of regularity which

mark off for us those happenings in the world around us which do require explanation by contrasting them with 'the natural course of events'—that is, those events which do not.[34]

Newton's first law is such an ideal. It specifies that uniform rectilinear motion is inertial motion, and that it is only changes in such motion that need to be explained. Newton's ideal of natural order displaced a corresponding Aristotelian ideal. Aristotle had taken as the paradigm case of local motion the dragging of a body over a resisting surface. The speed reached by such a body depends on the ratio of the effort exerted to the resistance offered. The very

[*] See Ch. 9, pp. 131–3.   [†] See Ch. 12, pp. 200–1.

presence of motion indicates that an effort is being applied. On the Aristotelian ideal of natural order, it is motion itself that needs to be explained and not just changes of motion. The two ideals conflict, and the triumph of the Newtonian ideal is a repudiation, and not an incorporation, of the Aristotelian ideal.

Toulmin declared that

an explanation, to be acceptable, must demonstrate that the happenings under investigation are special cases or complex combinations of our fundamental intelligible types.[35]

If a type of phenomena resists our best attempts to apply our principles of intelligibility, then it comes to be regarded as an anomaly. In the case of the Aristotelian ideal mentioned above, the motion of projectiles was an anomaly. On the Aristotelian ideal, the continued motion of a javelin after the hurler has released it, demands an explanation. But the airborne javelin appears to be subject to no effort. Aristotle suggested, with some hesitation, that the successively adjacent air transmits to the projectile a propensity to continue in motion.[36] Needless to say, Aristotelian natural philosophers were uneasy about interpretations of this type. Toulmin suggested that it is the recognition of anomalies which leads to the creation of new ideals of natural order.

Given a competition between ideals of natural order, it is the "fittest" that survive, "fitness" being a matter of conceptual integration and fertility. And because what is at stake in such a conflict is the adequacy of a conceptual innovation, the conflict cannot be resolved by an appeal to some "evidential calculus". Toulmin maintained that the logical reconstructionist programme for a logic of confirmation is of limited value, since such a logic is inapplicable to those important conflicts in which standards of intelligibility themselves are at issue.[37]

N. R. Hanson suggested that a conceptual revolution in science is analogous to a *gestalt*-shift in which the relevant facts come to be viewed in a new way.[38] Following Wittgenstein,[39] Hanson distinguished between 'seeing that' and 'seeing as'. Hanson emphasized that 'seeing as', the *gestalt* sense of seeing, has been important in the history of science.

Consider the sixteenth-century controversy about the motion of the earth. Suppose that Tycho Brahe and Kepler stand on a hill facing east at dawn. According to Hanson, there is a sense in which Tycho and Kepler see the same thing. They both "see" an orange disc between green and blue colour patches. But there also is a sense in which Tycho and Kepler do not see the same thing. Tycho "sees" the sun rising from below the fixed horizon. Kepler "sees" the horizon rolling beneath the stationary sun. To see the sun as Kepler sees it is to have effected a *gestalt*-shift.[40]

Feyerabend and Feigl on the Death of Orthodoxy

Feyerabend announced in 1970 that "philosophy of science" is "a subject with a great past".[41] Taken at face value, this is not a controversial claim. But Feyerabend meant to imply, as well, that "philosophy of science" is a subject without a future. The "philosophy of science" to which he referred was logical reconstructionism. He declared that

there exists an enterprise which is taken seriously by everyone in the business where simplicity, confirmation, empirical content are discussed by considering statements of the form (x) $(Ax \supset Bx)$ and their relation to statements of the form $Aa, Ab, Aa \& Ba$, and so on and *this* enterprise, I assert has nothing whatever to do with what goes on in the sciences.[42]

Feyerabend maintained that there is no reason for a practising scientist to consult the philosophy of science. There is nothing in the philosophy of science which can help him solve his problems. In particular, theories of confirmation do not help the scientist to decide which theories to accept. This is because theories of confirmation are based on two false assumptions. The first false assumption is that there is a theory-independent observation language with respect to which theories may be evaluated. The second false assumption is that it is possible for a theory to agree with all the known facts in its domain. But in practice there always is some evidence that counts against a theory. According to Feyerabend, it is as useless for a philosopher to base a theory of confirmation on this assumption as for a pharmaceutical house to produce a medicine which cures a patient only if he is free of all bacteria.

In Feyerabend's opinion, orthodox philosophy of science is a "degenerating problem-shift". Its practitioners ignore science in order to wrestle with problems about counterfactuals, 'grue', and confirmation. But all that this is good for is the generation of Ph.D. theses. The scientist is well advised to disregard it.

Nor is there any reason for a historian of science to study philosophy of science. There is nothing in orthodox philosophy of science which can help the historian to understand past progress in science.

Feyerabend's constructive proposal is to "return to the sources". The would-be philosopher of science should abandon the airy castles of logical reconstructionism and immerse himself in the history of science. Feyerabend praised the studies of specific episodes in the history of science made by Kuhn, Ronchi, Hanson, and Lakatos.[43]

"Return to the sources." No doubt this is good advice. But Feyerabend failed to make clear how a "philosophy of science" is implicated in, or is an outgrowth of, the history of science. Given a particular episode, what is it that a philosopher of science would do that distinguishes his enquiries from those of a historian of science?

Feyerabend doubtless would object that to pose such a question is to assume an inadmissibly parochial standpoint. Why should there be a distinct discipline—the philosophy of science—set apart from both the practice of science and the history of science? Indeed, why should there be a history of *science* distinct from the history of thought and action? Feyerabend is all for erasing the boundary lines drawn to separate "philosophy of science" from the broader pursuits of cultural history.[44] On his view, the philosophy of science is, and should be, an extinct discipline.

That is a pretty grim assessment. But then Feyerabend had made his reputation as a heretic. Herbert Feigl, by contrast, was unwilling to write off logical reconstructionism as a total loss.[45] Feigl had participated in the rise and reign of orthodoxy, and he looked back on its demise to see if orthodoxy had contained anything worth salvaging. He concluded that it had.

For one thing, the orthodox position explained how theories could be tested and compared. According to Feigl, the testing and comparing of theories is possible because

1. there are deductive relations between theories and empirical laws, and

2. there are numerous empirical laws which are "relatively stable and approximately accurate".

Of course, empirical laws are not incorrigible. In particular, they are subject to correction "from above". Feigl conceded that an astrophysical theory, for instance, may one day suggest revisions of its test-basis—the laws of physical optics. But he declared

I am not impressed with such purely speculative possibilities which the opponents of empiricism indefatigably keep inventing with shockingly abstruse super-sophistication! My point is very simply that thousands of physical and chemical ('low-level') constants figure in amazingly stable empirical laws.[46]

Feigl cited refractive indices, specific heats, thermal and electrical conductivities, and the regularities of chemical composition, as well as the laws of Ohm, Ampère, Coulomb, Faraday, Kirchhoff, and Balmer.

Feigl emphasized that he did not wish to claim that there is a theory-neutral observational language. Feigl suggested that the test basis for theories be shifted from observation reports to empirical laws. He declared that

while it may well be the case that all theories were (or are) 'born false'—i.e., that they all suffer from empirically demonstrable anomalies, there are thousands of empirical laws that—at least within a certain range of the relevant variables—have *not* required any revision or correction for decades,—some even for centuries of scientific development.[47]

The relative stability of empirical laws had been an important emphasis within orthodox philosophy of science. Ernest Nagel, for instance, had suggested that many laws have lives of their own which are independent of the theories advanced to explain them.[48]

Feyerabend had suggested that the meanings of the terms of an empirical law change as it is incorporated into successive high-level theories. Although its syntactic form may be unchanged in a transition, "the law" is different in each theory.

Feigl insisted that this emphasis on the theory-laden character of empirical laws fails to do justice to the role of laws in the practice of science. In practice, theories are appraised on their ability to account for empirical laws. On that score, Einstein's relativity theory is superior to Newton's mechanics, which, in turn, is superior to Galileo's theory of falling bodies. According to Feigl, orthodox theorists were correct to maintain that scientific progress is often an incorporation of laws into ever-more-inclusive theories.

Notes

1 Paul K. Feyerabend, 'An Attempt at a Realistic Interpretation of Experience' *Proc. Arist. Soc.* 58 (1958), 160–2.

2 Ibid. 164.

3 Peter Achinstein, *Concepts of Science* (Baltimore: The Johns Hopkins Press, 1968), 160–72.

4 Ibid. 168.

5 Pierre Duhem, *The Aim and Structure of Physical Theory* (New York: Atheneum, 1962), 180–218.

6 Willard van Orman Quine, 'Two Dogmas of Empiricism', in *From a Logical Point of View* (Cambridge, Mass.: Harvard University Press, 1953), 41.

7 Ibid. 43.

8 William Dray, *Laws and Explanation in History* (Oxford: Clarendon Press, 1957), 58–60.

9 Michael Scriven, 'Explanations, Predictions, and Laws', in H. Feigl and G. Maxwell (eds.), *Minnesota Studies in the Philosophy of Science*, iii, (Minneapolis: University of Minnesota Press, 1962), 170–230.

10 Richard Zaffron, 'Identity, Subsumption and Scientific Explanation', *J. Phil.* 68 (1971), 849–50.

11 Carl Hempel, 'Deductive–Nomological vs. Statistical Explanations', in Feigl and Maxwell (eds.), *Minnesota Studies in the Philosophy of Science*, iii. 109–10.

12 Hempel, *Aspects of Scientific Explanation* (New York: Free Press, 1965), 374–5.

13 Wesley Salmon, 'The Status of Prior Probabilities in Statistical Explanation', *Phil. Sci.* 32 (1961), 145.

14 Michael Scriven, 'Truisms as the Grounds for Historical Explanations', in P. Gardner (ed.), *Theories of History*, (Glencoe, Ill.: The Free Press, 1959), 443–75; 'Explanation and Prediction in Evolutionary Theory', *Science* 130, 1959, 447–82; 'Explanations, Predictions, and Laws', in Feigl and Maxwell (eds.), *Minnesota Studies in the Philosophy of Science*, iii. 170–230.

15 Hempel, *Aspects of Scientific Explanation*, 362.

16 Salmon, 'Why ask "Why"? An Inquiry Concerning Scientific Explanation', *Proc. Am. Phil. Soc.* 6 (1978), 689. Repr. in J. Kourany (ed.), *Scientific Knowledge* (Belmont, Calif.: Wadsworth, 1987), 56.

17 Frederick Suppe, 'The Search for Philosophic Understanding of Scientific Theories', in Suppe (ed.), *The Structure of Scientific Theories* (Urbana, Ill.: University of Illinois Press, 1974), 221–60.

18 Ibid. 222.

19 Pierre Duhem, *The Aim and Structure of Physical Theory*, trans. P. Wiener (New York: Atheneum, 1962), 32.

20 Wilfrid Sellars, 'The Language of Theories', in H. Feigl and G. Maxwell (eds.), *Current Issues in the Philosophy of Science* (New York: Holt, Rinehart, and Winston, 1961), 71–2; repr. in B. A. Brody (ed.), *Readings in the Philosophy of Science*, 348.

[21] Nelson Goodman, *Fact, Fiction and Forecast*, 2nd edn. (Indianapolis: The Bobbs-Merrill Co., Inc., 1965).

[22] Ibid. 74.

[23] Ibid. 78–80.

[24] Ibid. 94.

[25] Carl Hempel, 'Postscript (1964) on Confirmation', in *Aspects of Scientific Explanation* (New York: The Free Press, 1965), 51.

[26] Feyerabend, 'Explanation, Reduction, and Empiricism', in *Minnesota Studies in the Philosophy of Science*, iii. 46–8.

[27] Feyerabend, 'On the "Meaning" of Scientific Terms', *J. Phil.* 62 (1965), 267–71; 'Consolations for the Specialist', in I. Lakatos and A. Musgrave (eds.), *Criticism and the Growth of Knowledge* (Cambridge: Cambridge University Press, 1970), 220–1; 'Against Method: Outline of an Anarchistic Theory of Knowledge', in M. Radner and S. Winokur (eds.), *Minnesota Studies in the Philosophy of Science*, iv. 84.

[28] Feyerabend, 'On the "Meaning" of Scientific Terms', 271–2.

[29] Feyerabend, 'Explanation, Reduction, and Empiricism', 76–81.

[30] Hilary Putnam, 'How Not to Talk About Meaning', in R. Cohen and M. Wartofsky (eds.), *Boston Studies in the Philosophy of Science*, ii. (New York: Humanities Press, 1965), 206–7.

[31] Feyerabend, 'Reply to Criticism: Comments on Smart, Sellars and Putnam', in Cohen and Wartofsky (eds.), *Boston Studies in the Philosophy of Science*, ii. 229–30.

[32] Feyerabend, 'Explanation, Reduction, and Empiricism', 59.

[33] Stephen Toulmin, *Foresight and Understanding* (New York: Harper Torchbooks, 1961), 44–82.

[34] Ibid. 79.

[35] Ibid. 81.

[36] Aristotle, *Physics*, book VII, 267a.

[37] Toulmin, *Foresight and Understanding*, 112.

[38] N. R. Hanson, *Patterns of Discovery* (Cambridge: Cambridge University Press, 1958), ch. 4 and *passim*.

[39] Ludwig Wittgenstein, *Philosophical Investigation* (New York: Macmillan, 1953), 193–207.

[40] Hanson, *Patterns of Discovery*, 5–24.

[41] Feyerabend, 'Philosophy of Science: A Subject with a Great Past', in R. Stuewer (ed.), *Historical and Philosophical Perspectives of Science* (Minneapolis: University of Minnesota Press, 1970), 172–83.

[42] Ibid. 181.

[43] Ibid. 183.

[44] Feyerabend, *Against Method* (London: NLB, 1975), 294–309.

[45] Herbert Feigl, 'Empiricism at Bay?' in Cohen and Wartofsky (eds.), *Boston Studies in the Philosophy of Science*, xiv. 8.

[46] Ibid. 10.

[47] Ibid. 9.

[48] Ernest Nagel, *The Structure of Science*, 86–8.

14

Theories of Scientific Progress

THOMAS KUHN (1922–), received a Ph.D. in physics from Harvard, taught for many years at Princeton, and now is at MIT. He has contributed important historical studies of the Copernican Revolution and the origins of Quantum Mechanics. His widely influential book *The Structure of Scientific Revolutions* directed attention to the role of paradigms in the historical development of science.

IMRE LAKATOS (1922–74), a native of Hungary, was a victim of Nazi persecution who subsequently spent three years in jail during the era of Stalinist repression. In 1956 he left Hungary for England where he pursued investigations in philosophy of mathematics and philosophy of science at Cambridge and the London School of Economics.

LARRY LAUDAN (1941–), received a Ph.D. degree from Princeton. He has taught at Pittsburgh and VPI and is now at the University of Hawaii. Laudan has produced historical and critical studies of relationships among scientific theories, evaluative standards, and cognitive aims. His work is a valuable interpretation of the interdependence of philosophy of science and history of science.

Kuhn on "Normal Science" and "Revolutionary Science"

The numerous criticisms of orthodoxy had a cumulative effect. Many philosophers of science came to believe that something vital is lost when science is reconstructed in the categories of formal logic. It seemed to them that the proposed orthodox analyses of 'theory', 'confirmation', and 'reduction' bear little resemblance to actual scientific practice.

Thomas Kuhn's *The Structure of Scientific Revolutions* (first edition, 1962)[1] was a widely discussed alternative to the orthodox account of science. Kuhn formulated a "rational reconstruction" of scientific progress, a reconstruction based on his own interpretation of developments in the history of science. But Kuhn's reconstruction is not simply another history of science. Rather, it includes a second-order commentary—a philosophy of science—in which he presents normative conclusions about scientific method.

Toulmin and Hanson had indicated the direction that might be taken by a rational reconstruction of scientific progress. They had emphasized the importance of discontinuities in which scientists have come to see phenomena in new ways. Kuhn developed this emphasis into a model of scientific progress in which periods of "normal science" alternate with periods of "revolutionary science".

Normal Science

It is conceptual innovations which receive the most attention from historians of science. But much, if not most, science is carried on at a more prosaic level. It comprises "mopping-up operations"[2] in which an accepted "paradigm" is applied to new situations. Normal science involves

1. increasing the precision of agreement between observations and calculations based on the paradigm;
2. extending the scope of the paradigm to cover additional phenomena;
3. determining the values of universal constants;
4. formulating quantitative laws which further articulate the paradigm; and
5. deciding which alternative way of applying the paradigm to a new area of interest is most satisfactory.

Normal science is a conservative enterprise. Kuhn characterized it as "puzzle-solving activity".[3] The pursuit of normal science proceeds undisturbed so long as application of the paradigm satisfactorily explains the phenomena to which it is applied. But certain data may prove refractory. If scientists believe that the paradigm should fit the data in question, then confidence in the programme of normal science has been shaken. The type of phenomena described by the data is then regarded as an anomaly. Kuhn agreed with Toulmin that it is the occurrence of anomalies that provides the stimulus for the invention of alternative paradigms. Kuhn declared that

> normal science ultimately leads only to the recognition of anomalies and crises. And these are terminated, not by deliberation and interpretation but by a relatively sudden and unstructured event like the Gestalt switch.[4]

The competition between paradigms is quite unlike a competition between mathematical functions to fit a set of data. Competing paradigms are incommensurable. They reflect divergent conceptual orientations. Proponents of competing paradigms see certain types of phenomena in different ways. For example, where the Aristotelian "sees" the slow fall of a constrained body, the Newtonian "sees" the (nearly) isochronous motion of a pendulum.

Revolutionary Science

The presence of an anomaly or two is not sufficient to cause abandonment of a paradigm. Kuhn maintained that a logic of falsification is not applicable to the case of paradigm rejection. A paradigm is not rejected on the basis of a comparison of its consequences and empirical evidence. Rather paradigm-rejection is a three-term relation which involves an established paradigm, a rival paradigm, and the observational evidence.

Science enters a revolutionary stage with the emergence of a viable competing paradigm. It might seem that what is required at this stage is a comparison of the two paradigms and the results of observations. But such a comparison could be made only if there is available a paradigm-independent language in which to record the results of observations. Is such a language available? Kuhn thought not. He declared that in a sense that I am unable to explicate further, the proponents of competing paradigms practice their trades in different worlds. One contains constrained bodies that fall slowly, the other pendulums that repeat their motions again and again. In one, solutions are compounds, in the other mixtures. One is embedded in a flat, the other in a curved, matrix of space. Practicing in different worlds, the two groups of scientists see different things when they look from the same point in the same direction.[5]

Thus paradigm replacement resembles a *gestalt*-shift.[6] Competing paradigms are not wholly commensurable. Given a particular problem, two paradigms may differ with respect to the types of answer deemed permissible. For example, in the Cartesian tradition, to ask what forces are acting on a body is to ask for a specification of those other bodies that are exerting pressure on that body. But in the Newtonian tradition, one may answer the question about forces without discussing action-by-contact. It suffices to specify an appropriate mathematical function.[7] In addition, although a new paradigm usually incroporates concepts drawn from the old paradigm, these borrowed concepts often are used in novel ways. For instance, in the transition from Newtonian physics to General Relativity the terms 'space', 'time', and 'matter' undergo a far-reaching reinterpretation.[8]

Israel Scheffler complained that Kuhn's position on paradigm-replacement reduces the history of science to a mere succession of viewpoints.[9] Scheffler maintained that the history of science, unlike the history of philosophical systems, can be measured against a yardstick of descriptive adequacy. Progress in science can be measured because competing theories often make the same referential claims. To be sure, competing high-level theories may impose different systems of classification, but it often is the case that it is the same objects which are being classified.[10]

Scheffler suggested that Kuhn, by his appeal to the gestalt analogy, had promoted confusion between "seeing *x*" and "seeing *x*-as-something-or-other". Scheffler noted that it does not follow that because the classification systems of two paradigms differ, they are about different objects. It is possible that different paradigms introduce different ways of classifying one and the same set of objects.

Consider the case of the acceleration of electrons within a synchrotron. If we interpret the situation according to the conceptual scheme of Newtonian mechanics, then we attribute to the particles a "mass" that is independent of velocity. If we interpret the situation according to the conceptual scheme of Special Relativity Theory, then we attribute to the particles a "mass" whose value varies with velocity. The two concepts of "mass" are not the same. Nevertheless, if the competing theories are referentially equivalent in such applications, and if the Relativistic Interpretation achieves superior predictive success, then replacement of Newtonian Mechanics by Special Relativity Theory counts as progress.

Kuhn denied that competing paradigms can be measured against a yardstick of descriptive adequacy. However, he did insist that there are standards of rationality applicable to paradigm-replacement. Above all, a victorious paradigm must deal constructively with the anomalies that led to the crisis. And, other things being equal, a gain in quantitative precision counts in favour of a new paradigm.

In the first edition of *The Structure of Scientific Revolutions*, Kuhn specified a pattern of scientific progress to be superimposed on historical developments. Whether the pattern fits must be determined by historians of science. But before the historian can do this he must be clear about the outlines of the pattern. How is he to decide whether an experimental result is an anomaly, whether puzzle-solving activity has reached the crisis stage, or whether a *gestalt*-shift has occurred?

Unfortunately, Kuhn's usage of the concept of a 'paradigm' has been equivocal. Dudley Shapere[11] and Gerd Buchdahl[12] criticized Kuhn for shifting back and forth between a broad sense and a narrow sense of 'paradigm'.

In the broad sense, a "paradigm" is a "disciplinary matrix", or an "entire constellation of beliefs, values, techniques, and so on shared by members of a given community".[13] Members of a community of practitioners may share a commitment to the existence of theoretical entities (Absolute Space, atoms, fields, genes . . .). In addition, the members may be in agreement about which types of investigation and explanation are important (*in vivo v. in vitro* studies, contact-action *v.* field interpretations, deterministic *v.* probabilistic explanations . . .). Such commitments and beliefs are part of a "paradigm" in the broad sense. A disciplinary matrix also includes one or more "paradigms" in the narrow sense.

In the narrow sense, a "paradigm" is an "exemplar", an influential presentation of a scientific theory. Normally, exemplars are stated, augmented, and revised in textbooks which contain standard illustrations and applications of a theory.[14]

Shapere and Buchdahl pointed out the damaging effects of this equivocal use of "paradigm" on Kuhn's thesis about the history of science. If it is the narrow sense of "paradigm" that Kuhn has in mind, then the contrast between normal science and revolutionary science is greatly reduced. Instead of talking about "articulations of a single paradigm", the historian would have to discuss a succession of distinct exemplars. For instance in the narrow sense, Newton, d'Alembert, Lagrange, Hamilton, and Mach formulated different "paradigms" for mechanics. But transitions between such "paradigms" hardly merit the term "revolution". On the other hand, if it is the broad sense of "paradigm" that Kuhn has in mind, then the concept is too vague to be useful as a tool of historical analysis.

In a *Postscript* to the second edition of *The Structure of Scientific Revolutions* (1969), Kuhn conceded that his use of 'paradigm' had been equivocal.[15] He maintained, however, that historical-sociological inquiry may reveal both exemplars and disciplinary matrices. The sociologist first surveys conferences attended, journals read, articles published, literature cited, and the like. On the basis of this data, he identifies discrete "communities of practitioners". He then examines the behaviour of members of the community to see what commitments they share.

In his analysis of the likely outcome of such studies, Kuhn blurred the formerly sharp contrast between normal science and revolutionary science. He predicted that one result of sociological inquiry will be the identification of a large number of relatively small groups. He conceded that a revolution may occur within a micro-community without causing an upheaval within a science. He allowed for the replacement of one paradigm by another without the occurrence of a prior crisis within the micro-community. And he augmented the possible responses to a crisis situation to include the shelving of an anomaly for future consideration. But even more striking was Kuhn's concession that the pursuit of "normal science" within a micro-community may be accompanied by a debate over those metaphysical commitments that are basic to the "disciplinary matrix" of a science. He acknowledged that, in the nineteenth century, members of chemical communities pursued a common puzzle-solving activity in spite of differences of opinion about the existence of atoms. Members shared a commitment to the use of certain research techniques but disagreed, often vehemently, about the proper interpretation of these techniques.[16]

Several critics had complained that, in the first edition of *The Structure of Scientific Revolutions*, Kuhn had presented a caricature of science. Watkins, for instance, thought that Kuhn had depicted science as a series of widely spaced upheavals separated by lengthy dogmatic intervals.[17] However, in Kuhn's *Postscript*, normal science has lost whatever monolithic character it formerly had. Normal science is created by a micro-community in so far as its members agree on the research-value of an exemplar (paradigm). And Kuhn now allows for the replacement of an exemplar in the absence of any crisis. It would seem that Kuhn has disarmed the critics. Indeed, Alan Musgrave declared that "Kuhn's present view of 'normal science' will, it seems to me, cause scarcely a flutter among those who reacted violently against what they saw, or thought they saw, in his first edition."[18]

Lakatos on Scientific Research Programmes

The rational reconstruction of scientific progress was a much debated issue in the 1960s. Popper and Kuhn had provided the basic texts for the debate, and there followed a period of exposition and comparison. Perhaps the most important new standpoint to emerge from these discussions was that of Imre Lakatos.

Lakatos acknowledged that Kuhn was correct to emphasize continuity in science.[19] Scientists do continue to use theories in the face of evidence that seems to refute them. Newtonian mechanics is a case in point. Scientists in the nineteenth century recognized that the anomalous motion of Mercury counted against the theory. Nevertheless, they continued to use it. And they were not acting irrationally in so doing. Yet, according to Popper's methodological principles, it is irrational to ignore falsifying evidence. Lakatos criticized Popper for failing to distinguish between refutation and rejection.* Lakatos agreed with Kuhn that refutation neither is nor should be followed invariably by rejection. Theories should be allowed to flourish even within an "ocean of anomalies".

---

* Popper replied that Lakatos had misinterpreted him. Popper insisted that he had clearly distinguished the logical relation of refutation from the methodological question of rejection. He noted that the question of rejection depends, in part, on what alternative theories are available.[20]

But after awarding Kuhn high marks for his emphasis on continuity, Lakatos criticized him for treating revolutionary episodes as instances of "mystical conversion".[21] According to Lakatos, Kuhn has portrayed the history of science as an irrational succession of periods of rationality. This was most unfair to Kuhn. Although Kuhn did liken theory-replacement to the dawning of a new perspective, he did not maintain that scientific revolutions are irrational. I suppose that because "Kuhn-the-irrationalist" did not exist, it was necessary to invent him. "Kuhn-the-irrationalist" is a useful point of contrast for philosophers of science who believe that rules of appraisal can be found for theory-replacement.

Lakatos maintained that unless a rational reconstruction of theory-replacement can be given, the interpretation of scientific change must be left to historians and psychologists. Popper had produced a rational reconstruction, according to which scientific progress is a sequence of conjectures and attempted refutations. Lakatos sought to improve upon this reconstruction. In particular, he urged that the basic unit for appraisal should be "research programmes" rather than individual theories. According to Lakatos, a research programme consists of methodological rules: some tell us what paths of research to avoid (negative heuristic) and others what paths to pursue (positive heuristic).[22]

The negative heuristic of a research programme isolates a "hard core" of propositions which are not exposed to falsification. These propositions are accepted by convention and are deemed irrefutable by those who implement the research programme.

Examples of hard-core principles include:

Steno's Principle of Original Horizontality, a methodological principle for the interpretation of the geological column,

The Atomist Postulate that chemical reactions are the result of the association or dissociation of atoms, and

The Principle of Natural Selection.

The positive heuristic is a strategy for constructing a series of theories in such a manner that shortcomings at any particular stage can be overcome. The positive heuristic is a set of procedural suggestions for dealing with anticipated anomalies. As the research programme unfolds, a "protective belt" of auxiliary hypotheses is created around the hard core of non-falsifiable propositions.

Lakatos's Scientific Research Programme

For example, the Newtonian Research Programme[23] for the calculation of planetary and lunar orbits may be reconstructed as follows:

Theory	Auxiliary hypotheses	Results of applying theory
T_1	Sun stationary Sun and planet are point-masses such that $m_s \gg m_p$	Kepler's Laws deduced. Fit only approximate.
T_2	Sun and planet move about common centre of gravity	Improved fit, but motions of Jupiter and Saturn are anomalous.
T_3	Acknowledge perturbations Seek approximate solutions to 3-body interaction	Fit further improved. Anomalous motions of Jupiter and Saturn described by T_3. Motion of Moon anomalous.
T_4	Correction introduced for asymmetric mass-distribution	Motion of Moon described with improved accuracy by T_4. Anomalous motion of Uranus noted as more data becomes available.
T_5	A trans-Uranic planet exists	Neptune discovered near predicted location.

Tests of a research programme are directed at the protective belt of auxiliary hypotheses. Lakatos emphasized that a single negative test result does not refute an entire research programme. He criticized Popper for overestimating the importance of negative test results. Given a negative test result, a fruitful strategy may be to modify the protective belt of auxiliary hypotheses to accommodate the anomaly. And in some cases, the best available response may be to shelve the anomaly for future consideration.

But then how is a research programme to be appraised? Lakatos insisted, against Duhem and Kuhn, that there are rules of appraisal for sequences of theories. Some sequences constitute "progressive problem-shifts" and others constitute "degenerating problem-shifts".

A sequence of theories—T_1, T_2, ... T_n—is progressive if the following conditions are fulfilled:

(1) T_n accounts for the previous successes of T_{n-1};
(2) T_n has greater empirical content than T_{n-1}; and
(3) Some of the excess content of T_n has been corroborated.

Otherwise the problem-shift is degenerating.[24]

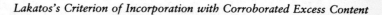

Lakatos's Criterion of Incorporation with Corroborated Excess Content

One way that a theory may "account for" the successes of its predecessor is by means of an asymptotic agreement of calculations. Thus historical episodes that satisfy Bohr's Correspondence Principle also satisfy the criterion 'incorporation with corroborated excess content'. Examples of theory-replacement that qualify include the transition from Ideal Gas Theory to van der Waals's Theory* and the transition from the Bohr Theory of the Hydrogen Atom (which restricts the electron to circular orbits) to the Bohr–Sommerfeld Theory (which permits elliptical orbits).

* See p. 144 above.

Lakatos emphasized that his criterion is an *objective* criterion. A research programme receives an affirmative evaluation only so long as it displays the power to anticipate and accommodate additional data.

However, this objective criterion must be applied at a particular time. And a research programme judged "degenerating" at a particular stage of its development may stage a comeback years later. Lakatos cited the changing fortunes of Prout's research programme, the aim of which was to show that the atomic weights of the chemical elements are exact multiples of the atomic weight of hydrogen (1.0 gm./gm. atom).[25] In 1816 the programme seemed promising. Further purification of samples of several elements led to atomic weight determinations that approached whole-number values. But the atomic weights of certain other elements, notably chlorine, remained fractional (Cl = 35.5 gm./gm. atom). Many chemists concluded that the Proutian programme was a degenerating problem-shift, and they abandoned it. Decades later, it was discovered that many elements occur in nature as mixtures of isotopes. In the case of chlorine, there are two isotopes—Cl^{35} and Cl^{37}. Newly developed techniques for separating isotopes were enlisted in the service of a revived Proutian programme.

Feyerabend complained that Lakatos's rules of appraisal are of practical value only when combined with a time limit. If no time limit is specified, then there is no reason ever to abandon a research programme. What seems at first to be a degenerating problem-shift may instead be the beginning stage of a long-term progressive problem-shift. As Feyerabend put it, "if you are permitted to wait, why not wait a little longer?"[26]

Lakatos replied that this objection is beside the point. Feyerabend has conflated two issues:

1. the methodological appraisal of a research programme, and
2. the decision whether to continue to apply a research programme.

With regard to the first issue, Lakatos called attention to the fact that he had specified rules of appraisal for research programmes. Admittedly the appraisal-verdict on a research programme may change with time. In particular, a negative experimental finding may come to be regarded as "crucial" against a programme only in retrospect.

With regard to the second issue, Lakatos insisted that it is not the duty of the philosopher of science to recommend research decisions to the scientist. Some scientists may choose to pursue a degenerating research programme in the hope that further work will re-establish the programme as progressive. Lakatos declared that "it is perfectly rational to play a risky game: what is irrational to deceive oneself about the risk".[27] To minimize opportunities for self-deception, Lakatos recommended that a cumulative public record be maintained of the successes and failures of each research programme.

Laudan on Problem-Solving

The work of Kuhn and Lakatos directed attention to the historical dimension of science. Much philosophy of science of the 1970s and 1980s sought to specify that which is progressive about science. Larry Laudan's *Progress and its Problems* (1977) was an important contribution to this project.

Laudan interpreted science to be a problem-solving activity. Thus the unit of progress within a scientific domain is the solved problem. According to Laudan, scientific problems can be subdivided into empirical problems and conceptual problems. Empirical problems are substantive questions about the structure or relations of domain-objects. Conceptual problems include problems that arise when incompatible or jointly implausible theories are entertained, or when there is incongruity between a theory and the methodological presuppositions of the domain. An example of the latter kind is the incongruity between the axiomatic structure of Newton's mechanics and Newton's professed inductivist theory of procedure. This conceptual incongruity was resolved only when certain of Newton's successors recognized that inductivism was not an adequate theory of procedure for theoretical physics. Conceptual problems are sometimes resolved by a change in methodological presuppositions. Thus the problem-solving model allows for evolving standards of rationality.

Progress is achieved within a domain when successive theories display increasing problem-solving effectiveness. Laudan sought to invert the logicist view of the relationship between rationality and progress. The logicist view is that developments in science are to be judged by an appeal to a standard of rationality. Developments that conform to the standard qualify as progressive. Laudan's position, by contrast, is that those developments which are progressive—which increase problem-solving effectiveness—qualify as rational.

Scientific progress may be achieved in a number of ways. One way is by an increase in the number of solved empirical problems. Laudan insisted that a theory may "solve" an empirical problem even if it entails only an approximate solution of the problem.[28] Thus Laudan would give credit to both Galileo and Newton for having solved the problem of free fall.*

A second type of progress is the resolution of an anomaly. Laudan took a broad view of anomalies. He held that an empirical result may count as an anomaly even if it is not inconsistent with the theory in question. This may happen, for instance, if a theory explains a particular result and its successor does not. For example, Descartes's Vortex Theory explained why the planets revolve around the sun in the same direction. Newton's theory of gravitational attraction did not. Some scientists maintained that this counted against Newton's theory. They were correct to do so. Laudan declared that

whenever an empirical problem, p, has been solved by any theory, then p thereafter constitutes an anomaly for every theory in the relevant domain which does not also solve p.[29]

An anomaly may be removed in several ways. The simplest way is by a revision of its empirical basis. Had the subsequently discovered planet Uranus displayed retrograde motion, the Newtonian theory would have been off the hook. A second way is to accommodate the anomaly by tacking on an auxiliary hypothesis. Newtonian theory, together with Laplace's Nebular Hypothesis, can account for the unidirectional motion of the planets. And a third way to remove an anomaly is by making significant changes in the relevant theory.

A third type of scientific progress is by a restoration of conceptual harmony among supposedly conflicting theories. Examples

---

* Galileo's solution is only approximately correct. Galileo stated that the acceleration of a body falling to the earth's surface is constant. But since the distance between a falling body and the centre of mass of the earth changes, so too does the gravitational force acting on the body and its acceleration.

include Clausius's demonstration that classical thermodynamics can be developed within the kinetic theory of gases,[30] and the research by Rutherford and others on energy production in radioactive decay, research which removed a seeming inconsistency between Kelvin's calculations of the age of the earth and Darwin's theory of evolution.[31]

Notes

[1] Thomas Kuhn, *The Structure of Scientific Revolutions*, 1st edn. (Chicago: University of Chicago Press, 1962).

[2] Ibid. 24.

[3] Ibid. 35–42.

[4] Ibid. 121.

[5] Ibid. 149.

[6] Ibid. 121.

[7] Ibid. 147.

[8] Ibid. 148.

[9] Israel Scheffler, *Science and Subjectivity* (New York: Bobbs-Merrill, 1967), 19.

[10] Ibid. 45–66.

[11] Dudley Shapere, 'The Structure of Scientific Revolutions', *Phil. Rev.* 73 (1964), 383–94.

[12] Gerd Buchdahl, 'A Revolution in Historiography of Science', *Hist. Sci.* 4 (1965), 55–69.

[13] Kuhn, *The Structure of Scientific Revolutions*, 1st edn., 175.

[14] Ibid. 43.

[15] Kuhn, 'Postscript-1969', in *The Structure of Scientific Revolutions*, 2nd edn. (Chicago: University of Chicago Press, 1970), 174–210.

[16] Ibid. 180–1.

[17] John Watkins, 'Against "Normal Science"' in I. Lakatos and R. Musgrave (eds.), *Criticism and the Growth of Knowledge* (Cambridge: Cambridge University Press, 1970), 31.

[18] Alan Musgrave, 'Kuhn's Second Thoughts', *Brit. J. Phil. Sci.* 22 (1971), 291.

[19] Imre Lakatos, 'Falsification and the Methodology of Scientific Research Programmes', in Lakatos and Musgrave (eds.), *Criticism and the Growth of Knowledge* (Cambridge: Cambridge University Press, 1970), 177.

[20] Karl Popper, 'Replies to My Critics', in *The Philosophy of Karl Popper*, ii, ed. P. Schilpp (La Salle, Ill.: Open Court, 1974) 1009.

[21] Lakatos, 'Criticism and the Methodology of Scientific Research Programmes', *Proc. Arist. Soc.* 69 (1968), 151.

[22] Lakatos, 'Falsification and the Methodology of Scientific Research Programmes', 132.

[23] Ibid. 135–6.

[24] Ibid. 116–18, 134.

[25] Ibid. 138–40.

[26] Paul Feyerabend, 'Consolations for the Specialist', in *Criticism and the Growth of Knowledge*, 215.

[27] Lakatos, 'History of Science and its Rational Reconstructions', in R. Buck and R. Cohen (eds.), *Boston Studies in the Philosophy of Science*, viii (Dordrecht: D. Reidel, 1971), 104 n.

[28] Larry Laudan, *Progress and Its Problems* (Berkeley, Calif.: University of California Press, 1977), 23–4.

[29] Ibid. 29.

[30] Ibid. 94–5.

[31] Joe D. Burchfield, *Lord Kelvin and the Age of the Earth* (New York: Science History Publications, 1975), 163–205.

15

Explanation, Causation, and Unification

WESLEY SALMON (1925—), developed an interest in probability and induction while studying at UCLA under Hans Reichenbach. Subsequently he has produced important work on the philosophy of space and time and patterns of scientific explanation. Salmon has taught at Indiana University and the University of Arizona, and currently is University Professor of Philosophy at Pittsburgh.

PETER RAILTON (1950—), received a Ph.D. degree from Princeton and currently teaches at the University of Michigan. He has published articles on moral theory, medical ethics, and value theory, as well as articles on scientific explanation and probability.

PHILIP KITCHER (1947—), received a Ph.D. degree from Princeton. He has taught at Vermont, Minnesota, and the University of California, San Diego. Kitcher has developed a Unification Theory of Explanation and has applied it in detail to explanatory contexts in biology. In addition, he has delivered an incisive criticism of the claim that Creationism is a viable scientific alternative to Theories of Organic Evolution.

Salmon's Causal Model

The Covering-Law Model makes no reference to causal relatedness. For that reason, the Deductive-Nomological Pattern (DN) is subject to the flagpole and barometer countercases, and the

Inductive-Statistical Pattern (IS) cannot account for Smith's leukemia.* Wesley Salmon suggested in essays dating from 1965[1] that effective scientific explanations specify causal mechanisms. In the case of Smith's illness, these mechanisms include the production of gamma rays upon nuclear fission, the modification of cellular structure by gamma rays, and the differential response of modified and unmodified cells to attack by the leukemia virus. To explain Smith's illness is to show how these causal mechanisms are statistically relevant to his misfortune.

Salmon took a "cause" to be an event that triggers a mechanism by which structure is produced and propagated. He unpacked the concept 'cause' by reference to the concepts 'process', 'intersection', and 'probability'. Following Bertrand Russell, he maintained that a process is the persistence of some entity, quality, or structure. Representative processes are the motions of bodies and the propagation of waves. Salmon maintained that processes may be subdivided into "causal processes" and "pseudo-processes". Causal processes transmit modifications, or "marks", impressed on them, pseudo-processes do not. The standard illustration of a pseudo-process is the illumination of the walls of a room by a rotating searchlight beam. Such a beam may be "marked" by placing a red filter along the 30 degree radius of the beam, but the redness of the light spot on the wall at this point is not transmitted as the beam continues to sweep out its path. Causal processes, by contrast, are the means by which structure is propagated from one region of space-time to another.[2]

According to Salmon, an adequate analysis of causation must account for both the *propagation* of structure and the *production* of structure. New structure is produced whenever two or more causal processes intersect in such a way that they undergo modifications that persist after the intersection.[3] (It is possible for causal processes to intersect without subsequent modification. An example is the intersection of two beams of light in which there are no photon-collisions.)[4]

Salmon distinguished two types of intersection in which the processes subsequently are modified—the "conjunctive fork" and the "interactive fork". In the conjunctive fork, the causal processes

* See pp. 209 above.

intersect in such a way that the production of a given effect does not alter the probability of other effects produced by that cause. The atomic bomb–leukemia correlation is an example of a conjunctive fork. The probability that an individual one mile from the epicentre will contract leukemia within ten years is independent of the probability that other individuals similarly situated will contract the disease. Let A and B be individual leukemia cases at the designated distance and C be the explosion. The joint probability that A and B occur, given C, is equal to the product of the individual probabilities, given C, viz.:

(1) $P[(A\&B)/C] = P(A/C) \times P(B/C)$

In a conjunctive fork, effects A and B each are produced by prior event C such that the following four conditions obtain:

(1) $P[(A\&B)/C] = P(A/C) \times P(B/C)$
(2) $P[(A\&B)/\bar{C}] = P(A/\bar{C}) \times P(B/\bar{C})$
(3) $P(A/C) > P(A/\bar{C})$
(4) $P(B/C) > P(B/\bar{C})$

Reichenbach has shown that these four conditions jointly imply

(5) $P(A\&B) > [P(A) \times P(B)]$[5]

viz. the probability of the joint occurrence of the two effects is higher than the product of the individual probabilities that the events occur. Presumably it is the occurrence of C that is responsible for this inequality. Nevertheless, the probabilities of A, given C, and B, given C, are independent. Thus the probability of the joint occurrence of A and B is greater than would be expected if the two events were statistically independent, and the statistical dependence itself arises from the causal relationship of the two events to common factor C.

Salmon recommended Reichenbach's "Principle of the Common Cause" as a valuable directive principle in the search for conjunctive forks.[6] The Principle of the Common Cause directs one to posit a common cause of events which occur with higher frequency than would be expected for independent occurrences.

Whereas the conjunctive fork characterizes the production of independent processes "arising under special background conditions", the interactive fork characterizes "direct physical interactions".[7] In an interactive fork the production of a given effect does alter the probability of other effects produced by that cause. Collision processes fit the pattern of the interactive fork. Consider the case of colliding billiard balls. Prior to impact the motion of a cue ball is a causal process whose structure is characterized by particular values of velocity, mass, and spin. After impact, the structure of this causal process has been altered, and the nature of the alteration depends on the type of collision that takes place. Given its initial motion, the probability that the cue ball will rebound at a 45 degree angle is linked to the probability that the eight ball will move in a certain way. If C is the motion of the cue ball before collision, A is its motion after collision, and B is the motion of the eight ball after collision, then

$$P[(A\&B)/C] > [P(A/C) \times P(B/C)].$$

A major advantage of Salmon's position on causal relatedness is the reconciliation thus achieved between two views of causation—the singularity view and the regularity view. In Salmon's view, a causal process is a singular occurrence. It is individual processes that propagate structure, and it is the intersection of individual processes that generates modifications of structure. But the conjunctive forks and interactive forks in which causal processes are involved are adequately characterized only in terms of statistical regularities.

Salmon's causal model of explanation fits well the behaviour of physical systems subject to classical dynamical and electromagnetic forces. But, as Salmon himself has recognized, straightforward causal analysis has not achieved success in quantum mechanics.

Quantum mechanics is an indeterministic theory. Suppose an atom of U^{238} emits an α-particle. A DN explanation may be given:

> All U^{238} nuclei have probability P of
>   emitting an α-particle during the interval
> Δt between t_0 and $t_0 + \Delta t$.
> Nucleus n is a U^{238} nucleus at t_0.
> _____
> ∴ Nucleus n has probability P of emitting an
> α-particle during Δt.

However, what is "explained" is not the emission of the α-particle but rather the probability of its emission during Δt.

A causal explanation is subject to the same limitation. The

appropriate causal mechanism is the tunnelling of a particle through the potential-energy barrier of a nucleus. Given the quantum theory of the tunnelling effect, one can calculate the probability that an α-particle is emitted by a nucleus of a given type within a specified time interval. However, no causal explanation is available of the fact that a particular α-particle was emitted at a particular time. Suppose that the α-particle that in fact was emitted strikes a photographic emulsion. To explain the change in the emulsion it is necessary to refer to an actual cause and not just a "probability of a cause".

Railton's Deductive–Nomological–Probabilistic Model

Peter Railton has developed a model of probabilistic explanation for such cases. The model includes three factors: (1) a DN argument for the probability of α-particle emission, (2) a causal account of the underlying mechanism for this probability, and (3) specific information about the actual fact of emission.[8]

This augmented explanatory model is not an argument. If it were taken to be an argument, it would be viciously circular, since clause (3) states that which is to be explained. Nevertheless, the explanatory account of the emission by atom n includes reference to the emission itself. Of what value, then, is this presumed explanation? According to Railton, what the augmented account accomplishes is to explain why a highly improbable chance event took place. Atom n emitted an α-particle within Δt because (1) there exists a finite, although low, possibility of emission during this interval, and (2) the atom did decay during the interval. Moreover, the augmented explanatory account asserts that the emission was a quantum-mechanical tunnelling through the potential energy barrier of atom n.

To those who would argue that this is no explanation at all, Railton's response is that this is the only type of explanatory account available for an indeterministic system. One cannot explain why n had to emit an α-particle during Δt. The emission was not necessary. Nor can one explain why it was probable that n emit an α-particle during Δt. It was not probable. What is left to be explained is that the decay occurred despite the fact that it was highly improbable.[9] It occurred because there is a small, but finite, probability of emission, a probability that is associated with the quantum-mechanical tunnelling effect, and as a matter of fact n did emit an α-particle during Δt.

Kitcher on Explanation as Unification

Philip Kitcher has complained that the search for a causal theory of explanation is misguided. According to Kitcher, 'causal related-ness' is to be unpacked by reference to 'explanatory success', and not *vice versa*. He maintained that

the 'because' of causation is always derivative from the 'because' of ex-planation. In learning to talk about causes or counterfactuals we are absorbing earlier generations' views about the structure of nature.[10]

Thus we make attributions of causal relatedness on the basis of prior acceptance of scientific explanations.

If this is correct, then it is important to find criteria to jus-tify transitions from one era of science to the next. It is 'theory-comparison', and ultimately 'scientific progress' that must be analysed.

Kitcher maintained that the appropriate criterion is comparative unification. Intuitively, unification is achieved within our store of scientific knowledge by 'minimizing the number of patterns of derivation employed and maximizing the number of conclusions generated'.[11] Of course, in many cases a trade-off may be required. A fully developed theory of comparative unification would set forth conditions under which a decrease in the number of patterns of explanation outweighs a loss in the number of conclusions generated, and conditions under which an increase in the number of conclusions generated outweighs a gain in the number of pat-terns of explanation. Emphasis on explanatory unification is in the tradition of Whewell. The concept 'explanatory unification', like the concept 'consilience', stipulates a set of conditions for justified theory replacement.

Kitcher sketched a unification model of scientific explanation as an alternative to causal models. In response, Salmon suggested that the two approaches express compatible and complementary aims of scientific explanation. The unification model articulates the aim of the systematization of empirical knowledge. The causal model articulates the aim of uncovering 'the hidden mechanisms by which nature works'.[12] We are dissatisfied with systematic theories that are divorced from all considerations of causal mechanism. And we are also dissatisfied with collections of causal relations that lack hierarchical organization.

The best scientific explanations are those that achieve unification in such a way that causal mechanisms are displayed. Nevertheless, a unification achieved without reference to causal mechanisms still counts as an explanatory success. And the uncovering of causal relations not integrated into a comprehensive theory also counts as an explanatory success.

Notes

[1] Salmon's contributions are reviewed in Wesley C. Salmon, 'Four Decades of Scientific Explanation', in Philip Kitcher and Wesley C. Salmon (eds.), *Scientific Explanation: Minnesota Studies in the Philosophy of Science*, xiii (Minneapolis: University of Minnesota Press, 1989), 3–219.

[2] Salmon, 'Why Ask "Why"? An Inquiry Concerning Scientific Explanation', *Proc. Am. Phil. Soc.* 6 (1978), 685–701. Repr. in J. Kourany (ed.), *Scientific Knowledge* (Belmont, Calif.: Wadsworth, 1987), 51–64.

[3] Salmon, 'Causality: Production and Propagation', *PSA 1980*, ed. P. D. Asquith and R. W. Giere (East Lansing, Mich.: Philosophy of Science Association, 1981), 60.

[4] Ibid. 60.

[5] Hans Reichenbach, *The Direction of Time* (Berkeley, Calif.: University of California Press, 1956), 160–1.

[6] Salmon, 'Why Ask "Why"?', 691–4; 'Causality: Production and Propagation', 54.

[7] Salmon, 'Causality: Production and Propagation', 62.

[8] Peter Railton, 'A Deductive–Nomological Model of Probabilistic Explanation', *Phil. Sci.* 45 (1978), 213–19.

[9] Ibid. 216.

[10] Philip Kitcher, 'Explanatory Unification and the Causal Structure of the World', in *Scientific Explanation*, ed. Kitcher and Salmon, 477.

[11] Ibid. 432.

[12] Salmon, 'Four Decades of Scientific Explanation', 182.

16

Confirmation and Evidential Support

CLARK GLYMOUR (1942–), studied under Wesley Salmon at Indiana University, and received a Ph.D. degree from that institution in 1969. He now teaches at Pittsburgh. Glymour's research interests have included inductive logic, theories of evidential support, convergent realism, and the use of computer programs to draw causal inferences from statistical correlations.

Bayesian Confirmation Theory

Goodman's "New Riddle of Induction" was an obstacle to the Logical Reconstructionist project of formulating a purely syntactical definition of qualitative confirmation. Goodman had shown that if hypothesis H receives support from evidence statement e, so also do alternative hypotheses H', H'', et al.*

Some philosophers of science concluded that the appropriate response to the "New Riddle" is to develop a *quantitative* theory that assigns a high degree of confirmation to H and low values to

---

* For instance, if H = 'all emeralds are green', then H' = 'all emeralds are grue', H'' = 'all emeralds are gred' ... See pp. 211–13 above.

the alternative "grue-type" hypotheses. One promising approach is to make use of the resources of probability theory.

Given the axioms of the probability calculus:

(1) $P(A) \geqq 0$, where A is a sentence within system S,
(2) $P(t) = 1$, for tautology t in S,
(3) $P(A \lor B) = P(A)$ & $P(B)$, for mutually inconsistent sentences A and B, and
(4) $P(A/B) = \dfrac{P(A\&B)}{P(B)}$, where $P(A/B)$ is the probability of A, given B, on the assumption that $P(B) > 0$,

it follows that:

$$P(A/B) = \frac{P(B/A)P(A)}{P(B)}. \quad *$$

This is "Bayes' Theorem".† The theorem may be adapted in the service of a theory of evidential support, such that

$$P(h/e) = \frac{P(e/h)P(h)}{P(e)}$$

where $P(h/e)$ is the probability conferred on hypothesis h by evidence e, and $P(h)$ is the "prior probability" of h independent of that evidence.

If there is a set of mutually exclusive and exhaustive hypotheses $h_1 \ldots h_n$,‡ then Bayes' Theorem takes the form

---

* Since

$$P(A\&B) = P(A/B)P(B) \text{ (from (4))}$$
$$P(B\&A) = P(B/A)P(A) \text{ (from (4))}$$

but

$$P(A\&B) = P(B\&A)$$

hence

$$P(A/B) = \frac{P(B/A)P(A)}{P(B)}$$

† A theorem of this form was proved by Thomas Bayes in 1763.[1]
‡ For h_n to be mutually exclusive and exhaustive,
  (1) h_i logically implies $\sim h_j$, for $i \neq j$, and
  (2) $P(h_1) \lor P(h_2) \lor \ldots P(h_n) = 1$.

$$P(h_1/e) = \frac{P(e/h_1)P(h_1)}{P(e/h_1)P(h_1) + P(e/h_2)P(h_2) + \ldots P(e/h_n)P(h_n)}$$

This relation seems to accord with certain of our intuitions. It seems natural to take

$$P(h_1/e) - P(h_1)$$

to be the degree of evidential support provided for hypothesis h_1 by additional evidence e. This support increases the more likely e is true given h_1, and decreases the more likely $(h_2 \lor h_3 \lor \ldots h_n)$ is true.

It follows from the above relation that

$$\frac{P(h/e)}{P(h^*/e)} = \frac{P(e/h)P(h)}{P(e/h^*)P(h^*)}$$

Suppose h is the hypothesis that all emeralds are green, h^* is the hypothesis that all emeralds are grue, and e is the report that an emerald examined prior to time t is green. Since the probability that an emerald examined before t is found to be green is the same on either hypothesis, the probability ratio is

$$\frac{P(h/e)}{P(h^*/e)} = \frac{P(h)}{P(h^*)}$$

On the Bayesian interpretation, the problem is to formulate a theory of evidential support which accords a higher prior probability to the "green hypothesis".

Before this can be accomplished, it is necessary to decide what is meant by "the probability of a hypothesis". And before this can be accomplished, it is necessary to decide how to interpret the term 'probability'. The principal options are:

1. the "frequency interpretation", which takes probabilities to be frequencies of occurrence of types of outcome in a long-run series of trials,

2. the "logical interpretation", which takes probabilities to be determined by logical relations between hypotheses and statements recording evidence, and

3. the "subjectivist interpretation", which takes probabilities to be measures of rational belief.

Most Bayesians affirm the subjectivist interpretation. It is far from clear, however, how degrees of rational belief are to be assigned. Consider scientists' appraisals of a newly formulated hypothesis. There may be considerable disagreement among them. Bayesian theorists acknowledge this. But they emphasize that such disagreement will narrow upon application of Bayes' Theorem to the accumulating evidence. Scientists who disagree about the prior probability of a hypothesis come to agree about its posterior probability.

The Bayesian learning process fits certain evaluative situations very well. Suppose balls are drawn from an urn that contains a mixture of white balls and black balls, and then replaced. There may be considerable disagreement about the prior probability that the first drawing will produce a white ball. After the first drawing, evidence about the colour of the ball drawn is used to calculate the posterior probability of drawing a white ball from the urn. The resultant posterior probability then becomes a prior probability for the second drawing, the result of which is used to calculate a further posterior probability. The initial disagreement over the probability of drawing a white ball from the urn is overcome gradually by repeated applications of Bayes' Theorem to the results of successive drawings.

Critics of the Bayesian approach have complained that the evaluation of scientific theories is not at all like the estimation of the percentage of white balls in an urn. Consider the following evaluative situation. Smith and Jones independently seek to confirm the newly formulated law of refraction. The law states that

$$\frac{\sin i}{\sin r} = k$$

where i is the angle of incidence and r is the angle of refraction of a light ray passing from medium 1 into medium 2, and k is a constant whose value depends on the nature of the two media.

Jones makes twenty separate measurements of the angle of refraction of light passing from air to water, in each case at an angle of incidence of 35 degrees. Smith, by contrast, makes measurements at five different angles of incidence for four different media-pairs. Scientists assign greater importance to the body of evidence accumulated by Smith. Other factors being equal, they prefer a

maximum of variety within the evidential support for a hypothesis.*
But this preference is not reflected in the Bayesian formula.

Some Bayesians have replied that this type of objection is misplaced.† They point out that Bayesianism is a theory of *inference*.
It seeks to measure the rational degree of belief to be accorded a
hypothesis upon acceptance of certain evidence. As such, Bayesian
theory ought not to be expected to provide guidance about what
degree of confidence is appropriate for specific experimental results.

The Problem of Old Evidence

Clark Glymour has emphasized that a theory of inference is not a
theory of scientific explanation. He complained that

particular *inferences* can almost always be brought into accord with the
Bayesian scheme by assigning degrees of belief more or less *ad hoc*, but we
learn nothing from this agreement. What we want is an explanation of
scientific argument; what the Bayesians give us is a theory of learning,
indeed a theory of personal learning.[3]

According to Glymour, an important shortcoming of the Bayesian
position is that it discounts evidence known to be true prior to
formulation of a theory. For old evidence e_o, $P(e_o/h) = P(e_o) = 1$.
In such a case $P(h/e_o) = P(h)$, and e_o does not increase the prior
probability of h. This is highly counterintuitive. Consider the following cases from the history of science:

Theories and Prior Evidence

Evidence	Theory
Precession of the equinoxes	Newtonian Gravitational Theory
No pores exist in the septum of the heart	Harvey's Theory of Circulation of the Blood
The weight of a calx is greater than the weight of its corresponding metal	Lavoisier's Oxygen Theory of Combustion
Null result of the Michelson–Morley experiment	Einstein's Special Relativity Theory
Anomalous perihelion of Mercury	Einstein's General Relativity Theory

---

* A related emphasis is Popper's distinction between instances that conform to the
requirements of a hypothesis and "serious tests" of the hypothesis. See pp. 179–81.

For each case, scientists at the time took evidence *e* to support Theory *T*. And most philosophers of science today agree with this assessment. Of course, if the above theories explained only the evidence in question, then the assessment would be different.

Bayesian theorists have responded to such complaints. Howson and Urbach, for instance, accommodate "antecedently known" evidence by reference to counterfactual degrees of belief. According to Howson and Urbach, when *e* is known to be true before *h* is formulated, $[P(h/e) - P(h)]$ measures the evidential support that *would be* provided by *e* just in case *e* *were* newly added to the stock of information relevant to one's rational belief in *h*.[4] Critics of this Bayesian response are sceptical that rules can be developed for the estimation of counterfactual degrees of belief.

Daniel Garber suggested a different resolution of the problem of prior evidence. According to Garber, what is achieved by the incorporation of old evidence by a hypothesis is knowledge that the hypothesis entails the evidence.[5] Hypothesis *h* receives support from prior evidence e_p provided that

$$P(h/e_p \& (h \to e_p)) > P(h/e_p)$$

The notation '$h \to e_p$' is a bit misleading. Hypothesis *h* does not itself imply e_p. Additional premisses are required—premisses that state relevant conditions, and often auxiliary hypotheses as well. For example, Newton's theory of gravitational attraction entails Kepler's Third Law on the assumption that several *non-interacting* point-masses revolve around a $1/R^2$ centre of force.

Once formulated, a theory gains evidential support from the fact that the entailment relation is newly recognized. This revised Bayesian position thus allows two types of increase in evidential support: new evidence that raises the posterior probability of a theory, and newly discovered entailment relations to old evidence.

Garber emphasized that evidential support is forthcoming in the latter case only if the entailment relation is discovered subsequently to formulation of the theory in question. If, on the other hand, a theory is formulated expressly to entail old evidence, then that evidence lends no support to the theory. Goodman has shown how

---

† (page 248) Among them, Howson and Urbach.[2]

indefinitely many hypotheses can be invented to entail a given body of evidence.*

Assessing the Impact of New Evidence

Richard W. Miller pointed out that there are two quite different types of response to the discovery of new evidence. One can apply the Bayesian formula to calculate a revised degree of belief in the hypothesis under examination. Alternatively, one can revise the relevant prior probabilities so that the degree of belief in the hypothesis remains unchanged.[6] For example, the creationist, confronted with data that show a close resemblance of island species to nearby mainland species, may revise his initial belief that such resemblance is implausible. The creationist

might conclude, contrary to his initial assumption, that the environments on islands and adjacent mainlands must be similar yet different in ways that make distinctive but similar species the most adaptive choice for a creative intelligence.[7]

Miller maintained that the Bayesian approach lacks a rule to determine when such *ad hoc* revision of prior probabilities is acceptable. He insisted that it will not do to stipulate that prior probabilities are inviolable. The history of science contains many episodes in which *ad hoc* revision of prior probabilities turned out to be fruitful. Darwin, for instance, sought to readjust expectations about what "ought be revealed" in the fossil record in response to the failure of paleontologists to uncover fossil remains of transitional forms.† Miller concluded that, since Bayesian Theory provides no help in deciding whether to readjust prior probabilities in the face of new evidence, it is inadequate as a theory of evidential support in scientific contexts.

---

* e.g. hypotheses about emeralds that are "grue", "grurple", "gred" . . . Each of these hypotheses has the required deductive relationship to the report of an emerald found to be green before time *t*.

† Other fruitful *ad hoc* responses include Copernicus's revision of expectations about stellar parallax, a phenomenon required by the heliostatic system but not observed at that time, and Galileo's revision of expectations about the telescopic magnification of heavenly bodies (the telescope removes the "adventitious rays" of stars so that apparent stellar diameters decrease when viewed telescopically).

Glymour on "Bootstrapping"

Clark Glymour suggested that scientific hypotheses sometimes receive evidential support by a process of "bootstrapping" in which one part of a theory is invoked in support of another.[8] Newton's *Principia* contains numerous examples of bootstrapping. Newton proved, for instance, that data on the motions of Jupiter's moons support the hypothesis of universal gravitational attraction. He did so by demonstrating that data on the moons' orbits, together with the first and second axioms of motion, imply that a $1/R^2$ force exists between the planet and each of its moons.

Glymour insisted that Newton achieved confirmation thereby, even though he used one part of his theory (e.g. $F = ma$) to support a second part of the theory (universal gravitational attraction). Glymour declared that

the central idea is that the hypotheses are confirmed with respect to a theory by a piece of evidence provided that, using the theory, we can deduce from the evidence an instance of the hypothesis, and the deduction is such that it does not guarantee that we would have gotten an instance of the hypothesis regardless of what the evidence might have been.[9]

In the above example, bootstrapping has been achieved because other force–distance correlations are consistent with the conjunction of the first and second axioms.

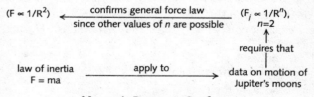

Newton's Bootstrap Confirmation

In another bootstrap application, Newton argued that the same force that accelerates bodies released near the surface of the earth also holds the moon in its orbit. The premises of the argument include the first and second axioms of motion, as well as data on falling bodies, the moon's orbit, and the earth–moon distance. Once again, Newton used one part of his theory to support another part of the theory.

Glymour did not claim that every case of evidential support fits the bootstrap model. It seems clear, however, that some important historical episodes do conform to the pattern.

Bootstrapping is achieved by deducing an instance of a hypothesis from the evidence, subject to certain restrictions. In so far as the Bootstrap Model takes confirmation to be a logical relation between sentences, it is in the Logical Reconstructionist tradition.

The logicist position on confirmation was expressed succinctly by Hempel in 1966:

> from a logical point of view, the support that a hypothesis receives from a given body of data should depend only on what the hypothesis asserts and what the data are.[10]

From this standpoint, the temporal relation between hypothesis and evidence does not matter. This temporal relation does matter, however, from the standpoint of historical theories of confirmation.

Lakatos on Comparative Confirmation

Goodman had shown that instances known prior to the formulation of a hypothesis (e.g. 'all emeralds are grue') may not confirm the hypothesis. Imre Lakatos undertook to specify conditions under which "old evidence", e_o, *does* provide support for hypothesis H. It does so, he concluded, provided that two conditions are satisfied:

1. H implies e_o,* and
2. there exists a competing "touchstone hypothesis" H_t such that either
   (a) H_t implies $\sim e_o$, or
   (b) H_t implies neither e_o nor $\sim e_o$.[11]

A touchstone hypothesis is a serious contender in the field, a contender that enjoys support from practising scientists.

Application of Lakatos's criterion requires historical inquiry. The philosopher of science must survey the scene to see whether there exist alternative hypotheses that do not imply the evidence. Old evidence provides support only within the context of competition between hypotheses.

---

* More accurately, H, in conjunction with statements about relevant conditions and suitable auxiliary hypotheses, implies e_o.

Thus Lakatos would hold that Lavoisier's Oxygen Theory of Combustion is confirmed by prior evidence on weight relations. Prior to Lavoisier's formulation of the Oxygen Theory, a number of studies had been made of the weight gained by metals upon combustion (e.g. by Boyle (1673), Lémery (1675), Freind (1709), and Guyton de Morveau (1770–2)).[12] This prior evidence was known to Lavoisier. Nevertheless, the data on weight relations confirm the Oxygen Theory because the data is inconsistent with the rival Phlogiston Theory.*

Notes

[1] Thomas Bayes, 'An Essay Towards Solving a Problem in the Doctrine of Chances', *Phil. Trans.* 53 (1763), 370–418. Repr. in *Biometrika*, 45 (1958), 296–315.

[2] Colin Howson and Peter Urbach, *Scientific Reasoning: The Bayesian Approach* (La Salle, Ill.: Open Court, 1989), 270–5.

[3] Clark Glymour, *Theory and Evidence* (Princeton, NJ: Princeton University Press, 1980), 74.

[4] Howson and Urbach, *Scientific Reasoning*, 270–5.

[5] Daniel Garber, 'Old Evidence and Logical Omniscience in Bayesian Confirmation Theory', in J. Earman (ed.), *Testing Scientific Theories* (Minneapolis: University of Minnesota Press, 1983), 99–131.

[6] Richard W. Miller, *Fact and Method* (Princeton, NJ: Princeton University Press, 1987), 297–319.

[7] Ibid. 315.

[8] Glymour, *Theory and Evidence*, 110–75.

[9] Ibid. 127.

[10] Carl Hempel, *Philosophy of Natural Science* (Englewood Cliffs, NJ: Prentice-Hall, 1966), 38.

[11] Imre Lakatos, 'Changes in the Problem of Inductive Logic', in I. Lakatos (ed.), *Inductive Logic* (Amsterdam: North-Holland, 1968), 376–7.

[12] See e.g. Henry Guerlac, *Lavoisier: The Crucial Year* (Ithaca, NY: Cornell University Press, 1967), 111–45.

* According to the Phlogiston Theory,

$$\text{metal} \xrightarrow{\text{heat}} \text{calx} + \text{phlogiston}$$
(calx + phlogiston)

Some Phlogiston-Theorists established consistency between their theory and the data by maintaining that the phlogiston liberated in combustion has a "negative weight".

17

The Justification of Evaluative Standards

In so far as different philosophies of science stipulate different
criteria of theory-replacement, they generate different reconstructions of scientific progess. Francis Bacon sees progress in science as
successive inductive generalization upon an expanding factual base.
Karl Popper sees progress as a sequence of bold, context-increasing
conjectures that survive attempts at refutation. And Imre Lakatos
sees progress as the articulation of scientific research programmes.
How are such competing rational reconstructions to be appraised?

Lakatos's Incorporation Criterion

Lakatos suggested that his own criterion of theory-replacement—
incorporation with corroborated excess context—is applicable as
well to sequences of methodologies.[1] He recommended the following
procedure for the evaluation of competing methodologies. First,
one selects a set of competing methodologies and elaborates the
rational reconstruction of scientific progress implied by each

methodology. Next one compares each rational reconstruction against the history of science. If methodology M_2 reconstructs all the historical episodes reconstructed by M_1, and additional episodes besides, then M_2 is the superior methodology.

Lakatos on the Evaluation of Competing Methodologies

Lakatos claimed that his own methodology of scientific research programmes is superior to Popper's methodology on this criterion. He noted that scientific research programmes are sometimes pursued in the face of dramatic falsifications. An example is the nineteenth-century pursuit of the Newtonian research programme in the face of anomalous data on the orbit of Mercury. Lakatos maintained that, on the Popperian reconstruction, such episodes are excluded from the rational growth of science. The methodology of scientific research programmes, by contrast, emphasizes the "relative autonomy of theoretical science",[2] and can account for the continued application of "refuted" principles.

Kuhn on the Circularity of Lakatos' Appraisal

In a review of Lakatos's position on the evaluation of rational reconstructions of scientific growth, Thomas Kuhn zeroed in on the apparent circularity of Lakatos's procedure.[3] Lakatos had maintained:

1. Philosophies of science imply rational reconstructions of scientific growth.
2. Each reconstruction delimits an "internal history" of science by marking off those episodes which fit its ideal of rationality from those episodes that do not (the "external history" of science).
3. The history of science can serve as a standard for the evaluation of rival methodologies. For instance, if more of the history of

science is rational under H_n than under H_{n-1}, then H_n is superior to H_{n-1}.

4. Every "history of science" is an *interpretation* of the historical record, an interpretation undertaken from a particular standpoint. The historian of science, *qua* historian, makes judgements of importance about the evidence available to him. These judgements of importance reflect his understanding of what counts as science and what kinds of factors affect its development.

However, if every history of science presupposes a methodological standpoint, there can be no methodologically neutral evaluation of historiographical theories. Lakatos judged the methodology of scientific research programmes to be superior to the falsificationist methodology on the basis of an appeal to a "history of science" formulated according to the canons of the methodology of scientific research programmes. The appraisal process is loaded in favour of the methodological commitments of the appraiser.

Kuhn's point is well taken. Lakatos's justificatory procedure does involve an element of circularity. The superior methodology is that methodology whose rational reconstruction of scientific progress best conforms to a history of science formulated according to the canons of that same methodology.

However, the circularity is open-ended. At a given point in time, M_3 may be superior to M_1 and M_2 when their respective rational reconstructions of scientific progress are compared to a history of science formulated according to the principles of M_3. Subsequently however, methodology M_4 may be formulated such that, given a history of science that reflects its assumptions, M_4 accounts for all the episodes accounted for by M_3, and additional episodes besides.

Laudan's "Standard-Case" Model

In *Progress and Its Problems* (1977), Laudan suggested an alternative procedure for the evaluation of competing methodologies.[4] This procedure avoids the circularity present in Lakatos's approach. It is anchored in a set of historical episodes deemed indisputably progressive by the "scientific élite" of the day. Competing methodologies are then evaluated on the basis of their abilities to reconstruct these "standard-case" episodes. The best methodology is the one that reconstructs the greatest number of the standard-case episodes. Once the best methodology is identified, it is used to formulate a history of science for episodes other than the standard cases.

Laudan on the Evaluation of Competing Methodologies

The results of Laudan's justificatory procedure are contingent upon the initial judgements of the scientific élite. These judgements are not subject to criticism. Laudan expressed confidence that the élite would agree upon a set of standard cases. His own candidates for inclusion in this set include

1. Newtonian Mechanics is superior to Aristotelian Mechanics on the evidence available in 1800;
2. the Kinetic Theory of Heat is superior to the Caloric Theory that takes heat to be a fluid on the evidence available in 1900, and
3. the General Theory of Relativity is superior to Newtonian Mechanics on the evidence available in 1925.[5]

Of course, the judgements rendered by the élite of the year 2050 may be quite different. If this turns out to be the case, then the methodology judged best today may be an "also ran" tomorrow. Laudan's procedure, like that of Lakatos, is open-ended. It is open-ended in two respects. New methodologies may be developed that capture more of the standard cases, and the standard cases themselves are subject to continuing review.

There is an additional problem. The best methodology is the one whose rational reconstruction captures the most standard-case episodes. But much depends on how the capture is achieved. A methodology may be augmented to reconstruct an additional standard-case episode by including evaluative principles tailored to account for just that episode. The added evaluative principles may be formulated in such a way that they apply only to the set of cultural conditions present at the time of the episode. Such an *ad hoc* adjustment presumably would not count as a "rational" reconstruction of the episode.

But to say this is to appeal to an antecedent understanding of 'rationality'. It would seem that circularity is present in Laudan's justificatory procedure as well. The justificatory procedure is supposed to single out the best set of evaluative principles for science. That methodology most congruent with the evaluative principles implicit in the standard-case episodes is taken to express the principles of scientific rationality. But to reach a decision about congruence it is necessary to invoke general principles of scientific rationality in order to block *ad hoc* moves of the above type.

The Sociological Turn

Lakatos and Laudan distinguished "internal history of science" from "external history of science". Internal history of science comprises those developments that can be reconstructed upon application of criteria of scientific rationality. External history of science comprises those episodes not subject to "rational reconstruction". Lakatos and Laudan conceded that social and political considerations may be invoked to account for external history of science. Presumably it is because of certain social or political pressures that external history of science fails to conform to the standards of scientific rationality. By contrast, there is a sense in which internal history of science is self-explanatory. To fulfil the conditions of scientific rationality is just to practice science as it should be done. No reference to extra-scientific social or political factors is required to explain why internal history of science developed as it did.

The division of labour implied by the internal–external distinction

was challenged during the 1970s and 1980s by a number of socio-
logists and philosophers. The centre of the protest was the University
of Edinburgh where David Bloor, Barry Barnes, and Steven Shapin
recommended a "Strong Programme" for the interpretation
of science.[6] At the core of the Strong Programme is a directive
principle—the interpreter of science is to uncover the causes of
scientists' beliefs by invoking the same types of cause to explain
both rational (true, successful) beliefs and irrational (false, unsuccessful) beliefs.[7] The aim of the Strong Programme is a causal
analysis that accounts for the development of both "internal history
of science" and "external history of science". Supporters of the
Strong Programme take the causes in question to be pressures
engendered by social structures. The Strong Programme thus inverts
the traditional pecking-order. The philosopher of science may
uncover some details about evaluative practice, but it is the sociologist who provides the important causal analyses of developments
in science.

Critics of the Strong Programme have complained that it ignores
the role of reasons in the formation of scientists' beliefs. Suppose a
scientist believes that a theory is true (probable, well-confirmed,
fertile). One type of cause of this belief is the further belief that
certain reasons are correct. Beliefs about reasons are often causes
of further beliefs.

In some cases, beliefs of scientists are caused by correct beliefs
about reasons. For example, Aristotle believed that Herodotus was
wrong to hold that female fish achieve conception by swallowing
the milt of a male. His stated reason for that belief is that the
passage from the mouth of a female fish is to its stomach and not
to its uterus.[8]

Of course, one can ask why Aristotle's realization that there is
no passageway from oesophagus to uterus led him to believe that
swallowing milt is irrelevant to conception. Perhaps there are social
forces that produce positive reinforcement in scientists who hold
beliefs for good reasons. But even so, a good explanation of
Aristotle's rejection of Herodotus' hypothesis is to cite Aristotle's
anatomical investigations of fish. For this specific belief, reference
to a reason carries more explanatory weight than does reference to
social factors.

A similar conclusion is appropriate in the case of Rutherford's
belief that atoms possess dense central nuclei. Rutherford supported

this belief by citing the results of scattering experiments. Most α-particles pass through a piece of gold foil without deflection, but an occasional particle is deflected through an angle of 90 degrees or more.[9]

The sociologist may discover social factors whose existence made it probable that Rutherford would perform α-particle scattering experiments. To the extent that the sociologist succeeds in this undertaking, she has added to our knowledge of this episode. But it remains a good explanation of Rutherford's belief in the nuclear atom that he believed that this model explained the facts of α-particle scattering.

It is clear that some beliefs of scientists are held because of further beliefs about reasons. Whether or not scientists' beliefs about reasons are themselves effects of social causes is an empirical question. The causal relevance of social factors must be argued on a case-by-case basis.

The Strong Programme is implausible. It is unlikely that an appeal to social factors can produce a sufficiently detailed causal account of theory-change.[10] Reference to social pressures may explain why certain *types* of theories were entertained or excluded (e.g. field theories rather than contact-action theories, deterministic theories rather than probabilistic theories), but reference to such pressures is unlikely to provide a complete causal account of the formation of a specific scientific theory. Of course this is an empirical question, but the burden of proof on defenders of the Strong Programme is heavy indeed.

Justification and Inviolable Principles

Lakatos and Laudan appraised competing methodologies on the assumption that there is a hierarchy of levels in the justification process.

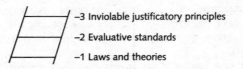

The Hierarchical Justificatory Ladder

Laws and theories are justified by appeal to standards of confirmation and explanation, and these standards are justified in turn, by appeal to trans-historical inviolable principles.

Level 3 is the "top rung" of the ladder. At level 3 Lakatos formulated an incorporation criterion, and Laudan developed a procedure which begins with a selection of standard-case episodes.

Dudley Shapere criticized this approach to justification. He denied that there exists a top rung on the ladder, a rung containing unalterable principles not themselves subject to justification. Rather, evaluative principles at all levels have been, and should be, subject to criticism and change. This applies to standards of evidential support, criteria of theory-replacement, interpretations of progress, and assumptions about the cognitive aims of science. Shapere recommended a "non-presuppositionist" philosophy of science according to which that enterprise involves no unalterable assumptions whatever, whether in the form of substantive beliefs, methods, rules, or concepts.[11]

Shapere maintained that transitions from one evaluative standard to another are often rational.[12] It is the task of a non-presuppositionist philosophy of science to exhibit this rationality. However, Shapere insisted that standards of rationality themselves change over time. Hence evaluative judgements are context-dependent. The philosopher may show that a transition from standard S_1 at time t_1 to S_2, at t_2 is rational, given the standards of rationality accepted at t_2. But this judgement may be incorrect, given the standards of rationality of some subsequent time. Since no suprahistorical standpoint is available for the appraisal of standards of rationality, non-presuppositionist philosophy of science is a version of historical relativism.

In *Science and Values* (1984), Laudan repudiated the hierarchial model of justification. He now agreed with Shapere that every evaluative level is subject to change. There is no unalterable "top rung". Indeed the "ladder model" is misleading. Laudan recommended instead a "reticulational model" within which theories, methodological principles, and cognitive aims are reciprocally interrelated.[13]

Laudan emphasized that justification is a two-way street. He noted that disputes about scientific theories often involve appeal to methodological principles. However, methodological principles

Laudan's Reticulational Model of Justification

themselves are sometimes changed in response to the success of substantive theories.

There is a similar reciprocal relation between theories and "axiological" claims about the basic cognitive aims of science. Shapere was correct to insist that even the cognitive aims of science are subject to change. Laudan noted, for example, that there was a tension within late eighteenth-century science between an acknowledged goal of Newtonian "Experimental Philosophy"—include within science only theories that correlate "manifest qualities"—and a proliferation of theories about unobserved entities.* According to Laudan, this tension was resolved in the nineteenth century by a revision of the axiological level to legitimate the invention of theories about entities not subject to observation.[14]

Laudan claimed that the reticulational model is superior to both the hierarchial model and "Kuhnian holism". "Kuhnian holism" is the position that theories, methodological rules, and cognitive aims are often replaced as a package. Before a revolutionary episode, scientists accept theories T, methodological rules M, and cognitive aims A. After the revolution, scientists accept T', M', and A'. The disciplinary matrix ('paradigm' in the broad sense) now is quite different. The holistic model promotes evaluative relativism. Before the revolution, theories are appraised by reference to M and A; after the revolution theories are appraised by reference to M' and

---

* Among the theories which posited unobserved entities were the phlogiston theory which interpreted combustion to be a process in which an invisible substance is emitted by the burning material, Hartley's neurological theory about the action of ethereal fluids, Franklin's one-fluid theory of electricity, and Lesage's theory of gravitational corpuscles.

A'. The transition $(T, M, A) \rightarrow (T', M', A')$ is not itself subject to justification. Any attempt to justify the revolution by appeal to M' or A' would be circular.

The reticulational model, by contrast, permits gradual, piecemeal adjustments among theories, methodological rules, and cognitive aims. Laudan sought to show that such adjustments are rational despite the fact that no component theory, rule, or aim is unalterable.

Laudan's reticulational model became the subject of a debate over the role of inviolable principles in the philosophy of science. Laudan had maintained that it is sometimes rational to resolve tensions by modifying the cognitive aims of science.

Gerald Doppelt complained that the reticulational model does not specify conditions under which it is rational to do so.[15] Laudan replied that there are two constraints on cognitive aims: they must be realizable* and they must be consistent with the values that inform the choice of theories.[16]

Doppelt pointed out that, if the cognitive aims of science are inconsistent with the values implicated in theory preference, then harmony may be restored either by changing aims or by changing theories.[17] If Laudan is correct that a shared aim of nineteenth-century scientists was to restrict theories to relations among "manifest qualities" then the disharmony introduced by theories about unobserved entities could have been removed by abandoning these theories. On the reticulational model, this too would be a rational response to disharmony.

Laudan conceded that the constraints on cognitive aims are relatively weak. Nevertheless he insisted that the constraints do provide an objective basis for evaluation, and that the reticulational model thereby avoids the relativism of Kuhnian holism.

In a review of *Science and Values*, John Worrall sought to revive the hierarchial model of justification. He declared that

If no principles of evaluation stay fixed, then there is no 'objective viewpoint' from which we can show that progress has occurred and we can say only that progress has occurred *relative to the standards that we happen to accept now*. However this may be dressed up, it is relativism.[18]

---

* It is unclear why unrealizability should disqualify a goal. It is not irrational for a historian to seek to write history "as it actually happened". It is not irrational for an engineer to strive to achieve a 100% foolproof spaceship launch procedure. And in the case of science, it is not irrational to strive for the complete reproducibility of experimental results.

Worrall suggested that the following evaluative principles be taken as inviolable:

1. Theories should be tested against plausible rivals (if there are any);[19]
2. Non *ad hoc* accounts should always be preferred to *ad hoc* ones (where both are available);[20] and
3. 'Greater empirical support can legitimately be claimed for the hypothesis that a particular factor caused some effect if the experiment testing the hypothesis has been shielded against other possible causal factors.'[21]

Worrall maintained that these methodological principles underlie evaluative practice in science just as *modus ponens* underlies deductive inference. A person who accepts p and $p \supset q$ but refuses to accept q has opted out of the game of deductive logic. Similarly, a person who rejects the basic scientific evaluative standards has opted out of the game of science. Rationality requires that the game be played according to these rules. Worrall insisted that ultimately we must stop arguing and 'dogmatically' assert certain basic principles of rationality.[22]

Laudan replied that Worrall's methodological rules are not purely formal rules like *modus ponens*.[23] For example, Rule 1 on testing a theory against plausible rivals is a *substantive* principle. There are possible worlds in which the principle would be counterproductive. In a world containing a finite number of ravens, each of which we have examined, it would be gratuitous to test the hypothesis 'all ravens are black' against alternative hypotheses. Laudan insisted that every substantive principle is subject to change with advancing knowledge.

According to Laudan, Worrall has misconstrued the nature of the threat from relativism. The threat is not that evaluative principles change (they do), but that there is no rationale for such change.[24] Laudan claimed that the reticulational model, subject to constraints of realizability and consistency, provides the needed rationale.

In response, Worrall conceded that no methodological principle is purely formal. In addition, he agreed that methodological principles have been created, modified, and abandoned within the history of science. But he denied that it follows from this that *every* such principle has been, or should be, subject to revision. Summarizing his dispute with Laudan, Worrall declared that

Laudan 'can see no grounds for holding any particular methodological rule—and certainly none with much punch or specificity to it—to be *in principle* immune from revision as we learn more about how to conduct inquiry'. Whereas it seems to me clear that in order to make sense of the claim that we '*learn* more' about how to conduct inquiry, some core evaluative principles must be taken as fixed.[25]

Notes

[1] Imre Lakatos, 'History of Science and Its Rational Reconstructions', in R. Buck and R. Cohen (eds.), *Boston Studies in the Philosophy of Science*, viii (Dordrecht: Reidel, 1971), 91–136.

[2] Lakatos, 'Falsification and the Methodology of Scientific Research Programmes', in I. Lakatos and A. Musgrave (eds.), *Criticism and the Growth of Knowledge*, (Cambridge: Cambridge University Press, 1970), 137; 'History of Science and Its Rational Reconstructions', 99.

[3] Thomas S. Kuhn, 'Notes on Lakatos', in Buck and Cohen (eds.), *Boston Studies in the Philosophy of Science*, vol. viii. 137–46.

[4] Larry Laudan, *Progress and Its Problems* (Berkeley, Calif.: University of California Press, 1977), 155–70.

[5] Ibid. 160.

[6] The Strong Programme is outlined by James R. Brown in the Introductory Essay in *Scientific Rationality: The Sociological Turn* (Dordrecht: Reidel, 1984), 3–40.

[7] See e.g. David Bloor, *Knowledge and Social Imagery* (London: Routledge & Kegan Paul, 1976), 5.

[8] Aristotle, *Generation of Animals*, 756^b.

[9] *The Collected Papers of Lord Rutherford of Nelson*, 4 vols., ed. J. Chadwick (New York: John Wiley, 1963), ii. 212–13; 423–31; 445–55.

[10] See e.g. Andrew Lugg, 'Two Historiographical Strategies: Ideas and Social Conditions in the History of Science', in *Scientific Rationality: The Sociological Turn*, 185–6.

[11] Dudley Shapere, 'The Character of Scientific Change', in Thomas Nickles (ed.), *Scientific Discovery, Logic and Rationality* (Dordrecht: Reidel, 1980), 94.

[12] Ibid. 68.

[13] Laudan, *Science and Values* (Berkeley, Calif.: University of California Press, 1984), 63.

[14] Ibid. 56–9.

[15] Gerald Doppelt, 'Relativism and the Reticulational Model of Scientific Rationality', *Synthèse*, 69 (1986), 234–7.

[16] Laudan, 'Relativism, Rationalism, and Reticulation', *Synthèse*, 71 (1987), 227–32.

[17] Doppelt, 'Relativism and the Reticulational Model of Scientific Rationality', 235.

[18] John Worrall, 'The Value of a Fixed Methodology', *Brit. J. Phil. Sci.* 39 (1988), 274.

[19] Ibid. 274.

[20] Worrall, 'Fix It and Be Damned: A Reply to Laudan', *Brit. J. Phil. Sci.* 40 (1989), 386.

[21] Ibid. 380.

[22] Ibid. 383.

[23] Laudan, 'If It Ain't Broke, Don't Fix It', *Brit. J. Phil. Sci.* 40 (1989), 373–4.

[24] Ibid. 369.

[25] Worrall, 'Fix It and Be Damned: A Reply to Laudan', 377.

18

The Debate over Scientific Realism

RICHARD BOYD (1942–   ), is Professor of Philosophy at Cornell University. He has published numerous articles in support of the position that the best explanation of the empirical success of science is that scientific theories are approximately true.

IAN HACKING (1936–   ), received a Ph.D. degree from Cambridge and currently teaches at the University of Toronto. He has argued in a number of publications that certain types of experimental manipulations by scientists provide support for the position of Entity Realism. Hacking also has written an influential history of theories about probability and inductive inference.

BAS C. VAN FRAASSEN (1941–   ), presently teaches at Princeton and serves as President of the Philosophy of Science Association. He has written on the Theory of Space and Time, Quantum Theory, and the role of symmetry considerations in theoretical science. Van Fraassen's Constructive Empiricism is a much-discussed alternative to Scientific Realism.

ARTHUR FINE (1937–   ), has developed a third alternative to the positions of Scientific Realism and Instrumentalism. He refers to this alternative as the "Natural Ontological Attitude" (NOA). In *The Shaky Game*, Fine presents NOA in the context of an analysis of the methodologies of Einstein, Schrödinger, and Bohr. Fine currently teaches at Northwestern University.

The realist–instrumentalist controversy flared anew in the 1970s. The controversy was a debate over

1. The proper cognitive aim of science, and
2. How best to account for the progress achieved within the history of science.

Truth Realism

The realist's answer to 1. is that scientists ought to seek to formulate true theories that depict the structure of the universe. The realist supports Galileo's position against instrumentalists like Pope Urban VIII who sought to restrict science to the "saving of appearances".

The realist's answer to 2. is that the record of progress indicates that the universe has a structure (largely) independent of human theorizing and that our theories have provided an increasingly more accurate picture of that structure. Realist philosophers of science in the 1970s called attention to the recent successes of Plate Tectonics Theory and the Theory of DNA Structure. It seemed clear that scientists had achieved new knowledge of the dynamics of geological change and heredity, and that these developments provided support for the realist position.

Hilary Putnam suggested in 1978 that unless one adopts a realist interpretation, the increasing predictive success achieved within the history of science would be a "miracle".[1] Putnam noted that the realism in question makes claims about both truth and existence. Within a scientific domain, increasing predictive success reflects an increasingly more adequate approximation to truth. And in so far as successive, predictively successful theories make different claims about particular theoretical objects (e.g. 'electrons', 'gravitational fields', 'genes'), these objects must exist.

Richard Boyd shifted attention from successful sequences of theories to the methodological principles implicit in the development of such sequences. Certain methodological principles are widely applied in the formulation of theories. Boyd argued that if predictively successful theories result, the best explanation of this success is a realist interpretation of the theories.[2]

One such principle is to formulate theories 'which quantify over familiar "theoretical entities"'.[3] Application of the principle presumably has given rise to theories of increasing instrumental reliability. Suppose a scientist develops theory T_2 by assigning an additional property or relation to the theoretical entities posited by T_1 (e.g. spin or elliptical orbits to the Bohr-Hydrogen-Atom electron). Suppose also that T_2 is superior to T_1 on grounds of predictive success. Boyd argued that the best explanation of this is that T_1 itself is approximately true, and that the approximation to truth has been improved by assigning a new property or relation to its theoretical entities.

Boyd put forward an "abductive" argument in support of this position:

(1) If successive theories within a scientific domain typically converge upon truth, then the principles of scientific method are instrumentally reliable.

(2) The principles of scientific method are instrumentally reliable (applications of the principles yield increasingly more reliable theories).

∴ It is probable that successive theories within a scientific domain typically converge upon truth.

Non-realists, by contrast, seek to uncouple the notions of predictive success and truth. Laudan, for instance, called attention to the long-term predictive success achieved by progressively modified Ptolemaic planetary models.[4] It is not because epicycle-deferent models are true of planetary motions that they achieved predictive success. Laudan emphasized that many scientific theories have achieved predictive success in spite of the fact that their central explanatory terms fail to refer. His list includes phlogiston theory, the caloric theory of heat, and the electromagnetic ether.[5] He concluded that successful prediction is not a reliable indicator of truth.

Laudan complained, in addition, that realists have failed to clarify what is meant by "approximate truth" or "progress toward truth". These concepts are parasitic upon the concept of truth.

Some scientific theories may be true. But in so far as they make universal claims they cannot be shown to be true. No amount of evidence can prove that unexamined instances resemble examined instances. Hume was correct about this.

But if we cannot show that a theory is true, how can we show that a sequence of theories constitutes progress toward truth? Laudan declared that

no one has been able even to say what it would mean to be 'closer to the truth', let alone to offer criteria for determining how we could assess such proximity.[6]

Entity Realism

The "convergence-upon-truth" thesis may be unconvincing. However, there are other ways to defend realism. In particular, one may argue that the entities posited by certain scientific theories do indeed exist. A strong case can be made for "entity realism" as opposed to "truth realism".

Rom Harré has analysed the claims of entity realism for three realms of cognitive objects. Within Realm 1, claims assert the existence of observable entities such as Mars, the Atlantic trench, and the renal portal vein.[7] Such claims can be adjudicated by appeal to reasonably straightforward experimental practices.

Within Realm 2, claims assert the existence of entities not presently observable. These existence claims arise within the context of "iconic theories". Iconic theories posit the existence of entities which, if real, are "objects of possible experience" subject to detection by the appropriately amplified human senses. For example, Harvey's Theory of Circulation is an iconic theory that posits the existence of connecting links between arteries and veins. According to the theory these presumed links are hollow vessels through which blood flows. When Malpighi discovered microscopic blood-carrying vessels connecting arteries and veins, he established that the objects posited by the theory indeed do exist. Micro-organisms and X-ray stars are additional Realm 2 cognitive objects whose existence has been established by subsequent instrumental evidence. Of course the conclusion that such entities exist is a conclusion based on theoretical considerations that pertain to the operation of scientific instruments.

The position of entity realism is that at least some of the cognitive objects discussed in scientific theories do exist. It suffices to establish this position that some entities in Realms 1 and 2 meet the existence criteria adopted within science.

Claims about Realm 3 entities are another matter. These claims assert the existence of entities which

if real, could not become phenomena for human observers, however well equipped with devices to amplify and extend the senses.[8]

Neutrinos are Realm 3 entities. They may be detected by reference to events they supposedly trigger, but they are not observable by the "amplified or extended" human senses.* It is an open question, best decided by practising scientists, whether satisfaction of a given type of detection procedure is an appropriate criterion of existence.

Quarks also are Realm 3 "entities". Quark-triplets are held to be associated with the transfer of basic forces within the nucleus. And yet, if current theory is correct, the isolated quark cannot exist.[9] Anti-realists are correct that quarks do not qualify as bona fide entities. However entity realism does not require that every theoretical object mentioned in an accepted scientific theory possess a referent.

Ian Hacking has emphasized that entity realism receives important support from facts about experimental inquiry. He noted that

entities that in principle cannot be observed are regularly manipulated to produce new phenomena and to investigate other aspects of nature.[10]

Consider the case of electrons. According to Hacking, our best evidence for the existence of electrons is not the explanatory power of theories about electrons, but experimental investigations in which electrons are manipulated to obtain information about other entities and processes. Among the important examples of such investigations are experiments conducted with the electron microscope. The electron microscope makes possible the determination of structures

---

* There is a very small probability that a neutrino, passing through a vat of dilute cadmium chloride solution, will interact with a hydrogen nucleus of a water molecule to produce a neutron and a positron. The positron is annihilated at once upon collision with an electron, producing two oppositely-directed γ-rays of 0.51 mev. energy apiece. The neutron travels a short distance before being absorbed by a cadmium ion. The capture is accompanied by a release of three or four γ-rays with total energy of 9 mev. Given an appropriate sequence of events—two 0.51 mv. oppositely-directed γ-rays followed shortly be three or four γ-rays of total energy 9 mev.—physicists conclude that a neutrino has struck a hydrogen nucleus, and hence that neutrinos exist. The rationale for this existence-claim is that no other known nuclear reactions produce precisely this configuration and sequence of γ-rays.

not visible through optical microscopes.* Our confidence that electrons exist is warranted because experimentalists have been able to enlist the causal properties of electrons to investigate "other more hypothetical parts of nature".[11] Similar considerations hold for theoretical entities other than electrons. A strong case can be made for the conclusion that many of the entities posited by scientific theories have been shown to exist.

Van Fraassen's Constructive Empiricism

The instrumentalist position is that scientific theories are calculating devices that facilitate the organization and prediction of statements about observations. It is the statements about observations that are true or false. Theories are merely "useful" or "not useful".

Bas van Fraassen's "constructive empiricism" is a variant of this position.[12] Van Fraassen introduced a distinction between truth and "empirical adequacy". Whereas "truth-realists" argue that scientific theories are true or false, van Fraassen insisted that the appropriate contrast is between theories that are empirically adequate and those that are not. An empirically adequate theory is a theory that is successful in saving the relevant phenomena. Van Fraassen maintained that the aim of science is to formulate empirically adequate theories and that it is no part of this aim to establish the truth of claims about theoretical entities.

Van Fraassen restricted beliefs about truth or falsity to statements that assign values to "observables". He accepted as "observables" only those concepts whose values can be determined by the *unaided* human senses. Thus a statement about craters on the surface of Neptune is a statement about observables, since it is empirically possible to decide its truth or falsity by direct observation (after an extensive journey). A statement about the motion of electrons, by contrast, is not a statement about observables, since such motions do not fall within the range of what can be observed by unaided human sense organs.

Certain statements about the motions of electrons are empirically adequate. For example, scientists regularly apply Relativity Theory

---

* e.g. scientists have used electron microscopes to produce images of the endoplasmic reticula of proteins.

and Quantum Theory to describe and predict motions within particle accelerators. Van Fraassen conceded that claims about the existence of theoretical entities are capable of being true or false. There is no dispute between Constructive Empiricism and Realism on the question of what a theory asserts. A claim such as "there exist electrons" is to be taken literally. However, van Fraassen recommended an agnostic attitude towards such claims. The constructive empiricist position is that empirical adequacy suffices for the purpose of science. Scientists are to restrict claims about truth and falsity to assertions about observables.

Elliott Sober complained that a scientist who follows this advice may be required to treat equivalent propositions differently. He called attention to the following sentences:

(1) There is a food web in which the human population occupies a terminal position.
(2) Human beings eat, but are not eaten by, other organisms.[13]

The constructive empiricist position is that it is appropriate to attribute a truth-value to (2) but not to (1), since 'food web' is not an observable. But (1) and (2) are roughly equivalent claims.

Ian Hacking opposed van Fraassen's restriction of observables to concepts whose values can be ascertained by our unaided sense organs. He called attention to the use of grids in the microscopic observation of small objects.[14] The grids are created by photographic reduction of hand-drawn intersecting lines. Often the squares of the macroscopic grid are lettered to facilitate the location of objects.

The macroscopic grid, and the microscopically-viewed photo-reduced grid, display the same pattern of labelled squares. Given this isomorphism,* it seems perverse to disqualify blood cells and other microscopically viewed objects from the realm of "observables". Indeed, scientists adjudicate existence claims by reference to what can be detected, and not solely by reference to what can be observed by unaided human senses. And as Hacking has emphasized, scientists are sometimes able to manipulate theoretical entities (e.g. electrons) in the investigation of other types of phenomena. Given the successful detection or manipulation of non-observable entities, it seems appropriate to believe that these entities do exist.

---

* Hacking noted that the isomorphism holds for a number of grid-manufacturing processes and a number of types of optical microscope.

Fine's Natural Ontological Attitude

Arthur Fine supported Hacking's conclusion. He noted that, within the context of specific research programmes, it is often fruitful to pose questions about truth and existence. To ask whether it is true that copper rods expand when heated, that giraffes have multi-chambered stomachs, or that electrons exist, is to ask questions that promote scientific progress.

Fine distinguished "local" and "global" appeals to realism.[15] Within science it is important to ascertain whether specific hypo-thesized entities exist. But realists and anti-realists often pose questions about "science-as-a-whole". In so doing, they assume that science is a set of practices in need of an interpretation.

The realist assumes that there exist entities so related as to constitute a structure that is (largely) independent of being observed. The aim of science is to formulate theories that represent this structure. Those theories that correspond to the structure of the world are true, and the global realist assumes that some theories do achieve the required correspondence (at least approximately).*

The global anti-realist, by contrast, denies that scientific theories can be shown to mirror the structure of the world. She maintains that it is predictive efficacy that counts, and that predictive success provides no warrant for claims about truth or existence.

Fine labelled his position the 'Natural Ontological Attitude".[16] The natural ontological attitude is to accept science as it is. This involves a commitment to accept the "certified results of science" as knowledge-claims on a par with the findings of common sense.[17] One may make this commitment without presupposing that specific scientific achievements are beyond doubt or that successive scientific interpretations are invariably progressive.

From the natural ontological perspective, claims about "the aim of science" resemble claims about "the meaning of life". In each case the appropriate strategy is to uncover the reasons why indi-viduals feel compelled to make global pronouncements, and then to provide appropriate therapy.[18] Fine maintained that

---

* Convergent realism is a variant of this position. The convergent realist maintains that successive theories may be shown to be increasingly better approximations to the truth.

the greatest virtue of NOA is to call attention to just how minimal an adequate philosophy of science can be.... For example, NOA helps us to see that realism differs from various anti-realisms in this way: realism adds an outer direction to NOA, that is, the external world and the correspondence relation of approximate truth; anti-realisms (typically) add an inner direction, that is, human-oriented reductions of truth, or concepts, or explanations.[19]

Fine insisted that the natural ontological attitude leaves open questions about the nature of truth. At any given time, there are established standards for assessing truth-claims within a domain of science, and it is the NOA position that questions about truth be examined by reference to these standards. Of course, the standards for judging truth are themselves subject to change with the growth of science. To adopt the natural ontological attitude is to accept this aspect of science as well.

Notes

[1] Hilary Putnam, 'What is Realism?', in Jarrett Leplin (ed.), *Scientific Realism* (Berkeley, Calif.: University of California Press, 1984), 140–1.

[2] Richard Boyd, 'The Current Status of Scientific Realism', in Leplin (ed.), *Scientific Realism*, 58–60; 'Scientific Realism and Naturalistic Epistemology', in *PSA 1980*, ii, ed. P. D. Asquith and R. N. Giere (East Lansing, Mich.: Philosophy of Science Association, 1981), 613–39.

[3] Boyd, 'Scientific Realism and Naturalistic Epistemology', 618.

[4] Larry Laudan, *Progress and Its Problems* (Berkeley, Calif.: University of California Press, 1977), 24, 46.

[5] Laudan, "A Confutation of Convergent Realism" *Phil. Sci.* 48 (1981), 33, reprinted in Leplin (ed.), *Scientific Realism*, 231.

[6] Laudan, *Progress and Its Problems*, 125–6.

[7] Rom Harré, *Varieties of Realism* (Oxford: Blackwell, 1986), 70–2.

[8] Ibid. 73.

[9] See Dudley Shapere, 'The Character of Scientific Change', in T. Nickles (ed.), *Scientific Discovery, Logic, and Rationality* (Dordrecht: Reidel, 1980), 72–3.

[10] Ian Hacking, 'Experimentation and Scientific Realism', in Leplin (ed.), *Scientific Realism*, 154.

[11] Ibid. 161.

[12] Bas van Fraassen, *The Scientific Image* (Oxford: Clarendon Press, 1980), 11–19.

[13] Elliott Sober, 'Constructive Empiricism and the Problem of Aboutness', *Brit. J. Phil. Sci.* 36 (1985), 16.

[14] Hacking, 'Do We See Through a Microscope?', *Pacific Phil. Quart.* 62 (1981), 305–22.

[15] Arthur Fine, 'The Natural Ontological Attitude', in Leplin (ed.), *Scientific*

Realism, 83–107; 'And Not Anti-Realism, Either', *Nous*, 18 (1984), 51–65, repr. in J. Kourany (ed.), *Scientific Knowledge* (Belmont, Calif.: Wadsworth, 1987), 359–68.

[16] Fine, 'The Natural Ontological Attitude', 97–102; 'And Not Anti-Realism, Either', in Kourany (ed.), *Scientific Knowledge*, 365–8.

[17] Fine, 'The Natural Ontological Attitude', 96.

[18] Fine, 'And Not Anti-Realism, Either', in Kourany (ed.), *Scientific Knowledge*, 366.

[19] Fine, 'The Natural Ontological Attitude', 101.

19

Descriptive Philosophies of Science

GERALD HOLTON (1922–), is Mallinckrodt Professor of Physics and Professor of the History of Science at Harvard. He has written extensively on the work of Kepler, Bohr, and Einstein. In his writings on the history of science, Holton has emphasized the role of thematic principles in the methodological decisions and evaluative decisions of scientists.

DAVID HULL (1935–), is Professor of Philosophy at Northwestern University and is a former president of the Philosophy of Science Association. Hull has studied research groups of scientists in action. In particular, he has traced the achievements and failures of "numerical taxonomists" and "cladists". At a more general level, Hull has developed a theory of science based on the interpretative categories of the Theory of Organic Evolution.

Philosophers of science from Aristotle to Kuhn have sought to develop evaluative standards applicable to the practice of science. These efforts have been informed by a prescriptive intent.* Prescriptivist philosophers of science have recommended standards by which scientific theories *ought* to be evaluated.

Some observers have found this prescriptivist project to be a bit presumptuous. The philosopher of science is in the position of seeking to educate scientists about proper evaluative practice. Of course, the philosopher may appeal to actual science "at its best"

---

* Fine's "natural ontological attitude" is an exception.

as source or warrant of the recommended standards. Nevertheless, prescriptivist philosophy of science is a legislative activity. And application of the prescribed evaluative standards is held to contribute to the creation of "good science".

Holton on Thematic Principles

Gerald Holton reported in 1984 that the theoretical physicists of the day appeared to have little interest in the recommendations of philosophers of science. He declared that it is the

perception by the large majority of scientists, right or wrong, that the messages of more recent philosophers, who themselves were not active scientists, are essentially impotent in use, and therefore may be safely neglected.[1]

Holton contrasted this supposed indifference with the intense interest in philosophical problems of such earlier scientists as Bohr, Einstein, and Bridgman.

Holton's response to this perceived change of attitude was to develop a purely descriptive philosophy of science. The descriptivist makes no recommendations about "proper" evaluative practice. Instead, he seeks to uncover the methodological standards and procedures that actually have informed scientific practice. These standards and procedures may or may not be the ones affirmed explicitly by the scientists in question. In the cases of Newton and Darwin for example, the descriptivist needs to distinguish their methodological practice from their pronouncements about this practice.

The descriptivist philosopher of science takes seriously Feyerabend's admonition, "return to the sources". Holton has contributed studies of Einstein, Millikan, Bohr, Kepler, Mach, and Stephen Weinberg, among others.[2] He concluded from these studies that certain thematic principles have been important in the historical development of science. These principles express scientists' basic commitments about the context of discovery and the context of justification. Thematic principles include:

1. explanatory principles (e.g. the "Ionian Enchantment", the ideal of a unified theory of all phenomena; Bohr's Principle of Complementarity);

2. directive principles (e.g. seek qualities within natural phenomena that are conserved, maximized, or minimized; seek to interpret macroscopic phenomena by reference to theories of micro-structure);
3. evaluative standards (e.g. parsimony, simplicity, incorporation);
4. ontological assumptions (e.g. atomism, plenism); and
5. high-level substantive hypotheses (e.g. quantization of energy, discreteness of electric charge, constancy of the velocity of light).[3]

Thematic principles are not written in stone. They are subject to modification (conservation of mass) and even abandonment (conservation of parity). Nevertheless, their influence is pervasive within the history of science. Indeed, Holton attributed the identity-through-time of the scientific enterprise to scientists' shared commitments to thematic principles. It is because there is widespread agreement about the types of theories to be developed and the types of explanation to be sought that science exists as a co-operative cumulative enterprise.

This is not to say that every appeal to a thematic principle is successful. Upon occasion, commitments to thematic principles have led scientists to overlook considerations which, in retrospect, ought to have been taken into account. Nevertheless, a second-order commentary on scientific evaluative practice would be incomplete without an examination of the pervasive influence of these principles.

To interpret science it is necessary to adopt some interpretative categories. William Whewell had emphasized this, maintaining that scientific theorizing involves a relationship between facts and ideas and that there are individual sciences, each with a set of fundamental ideas and first principles.*

In like manner, Holton recommended an interpretative framework for descriptive philosophy of science—the activities of scientists are to be set within a three-dimensional grid whose axes represent empirical content, analytical content, and thematic content.

Holton's distinctive contribution was to emphasize the role of thematic principles in the development of science. It is only by reference to the thematic axis that one can formulate plausible answers to the questions overleaf:

* See Ch. 9.

Holton's Three-Dimensional Grid

1. What is constant in the ever-shifting theory and practice of science—what makes it one continuing enterprise, despite the apparently radical changes of detail and focus of attention?
2. Why do scientists, at enormous risk, hold on to a model of explanation, or to some 'sacred' principle, when it is in fact being contradicted by current experimental evidence? [and]
3. Why do scientists . . . with good access to the same information often come to hold so fundamentally different models of explanation?[4]

Holton did not issue prescriptive recommendations on behalf of specific thematic principles; in that respect his approach is descriptive. His sole prescriptive claim is that an adequate philosophy of science must analyse methodological and valuational practice by reference to an interpretative framework sensitive to the influence of thematic principles.

Toulmin on Conceptual Evolution

Stephen Toulmin has put forward an alternative model for descriptive philosophy of science. Toulmin's model is biological rather than geometrical. It is an application of the Darwinian Theory of Evolution to the historical development of science.

Toulmin recommended that philosophers of science shift attention from logical relations between propositions to the progressive modification of concepts. He maintained that important questions in science often take the form:

given that concepts c_1, c_2,..., are in some respect *inadequate* to the explanatory needs of the discipline, how can we modify/extend/restrict/qualify them, so as to give us the means of asking more fruitful empirical or mathematical questions in this domain?[5]

Toulmin held that conceptual development is an "evolution" within which "natural selection" operates on a set of "conceptual variants". It is the "fittest" concepts that survive.[6]

Toulmin's evolutionary model meshes well with Kuhn's description of scientific revolutions. A revolution is a competition among paradigms (sets of concepts). It is the paradigm that achieves best adaptation to explanatory pressures within the discipline that wins the field. The victorious paradigm is the one that best resolves the anomalies (changed environmental conditions) that led to the revolutionary crisis.

L. J. Cohen pointed out that there is an important disanalogy between the development of science and the evolution of species. Conceptual evolution, unlike organic evolution, involves processes that are "coupled". There is a connection between the generation of conceptual variants and the selection of these variants. Conceptual variants are not "mutations" that arise in a spontaneous, random manner. Scientists invent conceptual variants to solve specific disciplinary problems. Thus, it is the judgements of scientists that are responsible for both the generation and the selection of concepts. By contrast, in organic evolution, mutation and selection are "uncoupled". As Cohen put it,

the gamete has no clairvoyant capacity to mutate preferentially in directions preadapted to the novel ecological demands which the adult organisms are going to encounter at some later time.[7]

Toulmin and Cohen agreed that this disanalogy exists, but disagreed about its importance. Cohen insisted that the lack of coupling between the generation of variants and the selection of variants is an essential feature of Darwinian theory, and that an interpretation of conceptual change that lacks this feature is not genuinely Darwinian. Toulmin held that evolutionary biology and conceptual change share enough points of similarity for the analogy to be useful for the interpretation of science.

Hull on Selection Processes

David Hull agreed with Toulmin. Hull developed a "General Theory of Selection Processes" which takes selection to be a

process in which the differential extinction and proliferation of interactors cause the differential perpetuation of the relevant replicators.[8]

Replicators are entities of which copies are made and transferred. In sexually reproducing organisms, replicators are usually genes. Interactors are entities subject to competition within an environment. Over time, the selection process gives rise to lineages. A lineage is

an entity that changes indefinitely through time either in the same or an altered state as a result of replication.[9]

A lineage is a sequence of replicators. It is also an individual, a temporally bounded segment of an evolutionary path.

Hull interpreted both biological evolution and the history of science to be selection processes. Within science, it is concepts that are replicators and individual scientists and research groups that are interactors.

The General Theory of Selection Processes provides a set of categories for the interpretation of the history of science. It is the "fittest" conceptual innovations that survive. Fitness is to be assessed with respect to "environmental pressure" within the social-institutional matrix of science.

Fitness in science, as in organic evolution, is a balance between adaptation to present conditions and retention of capacity to respond creatively to future changes of these conditions. Thus judgements about the success of particular conceptual changes are always provisional. It may be the case that a presently effective conceptual adjustment diminishes the future fertility (adaptability) of the relevant theory.

The interpreter of science who applies Hull's General Theory is concerned to trace lineages of concepts. What counts are *causal* relations in the evolutionary process and not questions about sameness of content. Hull noted, for instance, that the researches of Darwin and the researches of A. R. Wallace are both included in the lineage of natural selection theory, whereas the independent, but non-influential version of the theory by Patrick Matthew (1831) is not.[10] Hull insisted that only those conceptual innovations that

Hull's Theory of Selection Processes

	Biological Evolution	Theory of Selection Processes	History of Science
Units of variation	mutant forms within a population at time t_1	*replicators*—units of heredity of which copies are made	concepts, beliefs, investigative techniques
Units of effective modification	those t_1 variants that are dominant within the population at t_2	*interactors*—units involved in adaptive competition	individual scientists, research groups
Products of interaction	species	*lineages*—historical individuals (genealogical segments) and not classes	lineages of concepts
Mechanism	natural selection	differential perpetuation of replicators resulting from "genealogical actors performing in ecological plays"	scientists seeking credit constrained by checks

are acknowledged and utilized by subsequent investigators participate in lineages. It is genealogy and not structural similarity that is determinative.

Hull applied the General Theory of Selection Processes in two ways—as a framework for the interpretation of the history of science, and as a theory of science. As a theory of science, the General Theory provides answers to certain puzzling questions about the historical development of science. Among these questions are:

1. Why is science so successful in achieving its acknowledged aims?
2. Why, if it is the formulation of effective theories that matters, are scientists so concerned about matters of priority and proper citation? and
3. Why are the self-policing activities of science so effective, in contrast to the inept self-policing efforts of other professions?

Hull attributed the success of science to the fact that the self-interests of individual scientists coincide with the aims of the discipline.[11] That which best contributes to a scientist's career is the publication of work that is acknowledged and used by her fellow scientists.

It is the contribution that a scientist makes to the research success of her peers that establishes her "fitness" *qua* "interactor". It would be evolutionarily self-defeating to fabricate data or otherwise undermine the scientific enterprise. Hull's General Theory of Selection Processes explains why cases of professional misconduct are rare among scientists. That the Theory of Selection Processes provides a rationale for the success of science is a point in its favour as an interpretative framework for the delineation of conceptual lineages.

Descriptive Philosophy of Science and the History of Science

The descriptive version of the philosophy of science has the virtue of modesty. The philosopher is to be an exhibitor and not an advocate. Scientists are left free to apply, modify, or ignore the evaluative standards uncovered within the descriptive philosophy of science.

It might seem that the descriptive approach subsumes the philosophy of science under the history of science. The philosopher of science becomes a historian with a particular interest in evaluative practice. This is not quite correct. There remains an important difference of intent. Whereas the historian seeks to create explanatory narratives, the philosopher seeks to develop evaluative principles applicable to diverse instances. As Kuhn has emphasized, it is this interest in that which is general that sets the philosopher apart from the historian.[12] It remains to be seen whether the descriptive approach to the philosophy of science will flourish.

Notes

[1] Gerald Holton, 'Do Scientists Need a Philosophy?', *Times Literary Supplement*, 2 Nov. 1984, 1232.

[2] Holton, 'Thematic Presuppositions and the Direction of Scientific Advance', in A. F. Heath (ed.), *Scientific Explanation* (Oxford: Clarendon Press, 1981); *Thematic Origins of Scientific Thought*, rev. edn. (Cambridge, Mass.: Harvard University Press, 1988); *The Scientific Imagination* (Cambridge: Cambridge University Press, 1978).

[3] Holton, 'Thematic Presuppositions and the Direction of Scientific Advance', 17–23; *Thematic Origins of Scientific Thought*, 10–68; *The Scientific Imagination*, 6–22; 'Do Scientists Need a Philosophy?', 1235.

[4] Holton, *The Scientific Imagination*, 7.

[5] Stephen Toulmin, 'Rationality and Scientific Discovery', in K. Schaffner and R. Cohen (eds.), *Boston Studies in the Philosophy of Science*, xx (Dordrecht: D. Reidel, 1974), 394.

[6] Ibid. 394–406; *Human Understanding*, i (Oxford: Clarendon Press, 1972).

[7] L. Jonathan Cohen, 'Is the Progress of Science Evolutionary?' *Brit. J. Phil. Sci.* 24 (1973), 47.

[8] David L. Hull, *Science as a Process* (Chicago: University of Chicago Press, 1988), 409; *The Metaphysics of Evolution* (Albany, NY: SUNY Press, 1989), 96.

[9] Hull, *The Metaphysics of Evolution*, 106.

[10] Ibid. 233.

[11] Hull, *Science as a Process*, 303–12.

[12] Thomas S. Kuhn, 'The Relations Between the History and Philosophy of Science', in *The Essential Tension* (Chicago: University of Chicago Press, 1977), 3–20.

Select Bibliography

A *useful bibliography of sources for the history of the philosophy of science is*

LAUDAN, L., 'Theories of Scientific Method from Plato to Mach: A Bibliographical Review', *History of Science*, 7 (1969), 1–63.

1. Aristotle's Philosophy of Science

Works by Aristotle

Posterior Analytics, trans. with notes by J. Barnes (Oxford: Clarendon Press, 1975).

The Works of Aristotle Translated into English, ed. J. A. Smith and W. D. Ross, 12 vols. (Oxford: Clarendon Press, 1908–52).

Works about Aristotle

ALLAN, D. J., *The Philosophy of Aristotle*, 2nd edn. (London: Oxford University Press, 1970).

ANSCOMBE, G. E. M., 'Aristotle: The Search for Substance', in Anscombe and P. T. Geach, *Three Philosophers* (Oxford: Blackwell, 1961).

APOSTLE, H., *Aristotle's Philosophy of Mathematics* (Chicago: University of Chicago Press, 1952).

BARNES, J., SCHOFIELD, M., and SORABJI, R. (eds.), *Articles on Aristotle* i (London: Duckworth, 1975).

DEMOS, R., 'The Structure of Substance According to Aristotle', *Phil. and Phenom. Res.* 5 (1944–5), 255–68.

EVANS, M. G., 'Causality and Explanation in the Logic of Aristotle', *Phil. and Phenom. Res.* 19 (1958–9), 466–85.

FURTH, M., *Substance, Form and Psyche: An Aristotelean Metaphysics* (Cambridge: Cambridge University Press, 1988).

GOTTHELF, A., and LENNOX, J. (eds.), *Philosophical Issues in Aristotle's Biology* (Cambridge: Cambridge University Press, 1987).

GRAHAM, D. W., *Aristotle's Two Systems* (Oxford: Clarendon Press, 1987).

GRENE, M., *A Portrait of Aristotle* (Chicago: University of Chicago Press, 1963).

HALPER, E., 'Aristotle on Knowledge of Nature', *Rev. Meta.* 37 (1984), 811–35.

IRWIN, T., *Aristotle's First Principles* (Oxford: Clarendon Press, 1988).

LEAR, J., *Aristotle and the Desire to Understand* (Cambridge: Cambridge University Press, 1988).

LEE, H. D. P., 'Geometrical Methods and Aristotle's Account of First Principles', *Class. Quart.* 29 (1935), 113–24.

McKEON, R. P., 'Aristotle's Conception of the Development and the Nature of Scientific Method', *J. Hist. Ideas* 8 (1947), 3–44.

MATTHEN, M. (ed.), *Aristotle Today: Essays on Aristotle's Ideal of Science* (Edmonton: Academic Printing and Publishing, 1986). See particularly the essays by M. Matthen, F. Sparshott, and M. Furth.

RANDALL, J. H., Jun., *Aristotle* (New York: Columbia University Press, 1960).

ROSS, W. E., *Aristotle*, 5th edn., rev. (London: Methuen, 1949).

SELLARS, W., 'Substance and Form in Aristotle', *J. Phil.* 54 (1957), 688–99.

SOLMSEN, F., *Aristotle's System of the Physical World* (Ithaca, NY: Cornell University Press, 1960).

2. The Pythagorean Orientation

CORNFORD, F. M., *Plato's Cosmology* (New York: Liberal Arts Press, 1957), a translation of Plato's *Timaeus* with running commentary by Cornford.

GUTHRIE, W. K. C., *A History of Greek Philosophy*, i (Cambridge: Cambridge University Press, 1962).

HARRÉ, R., *The Anticipation of Nature* (London: Hutchinson, 1965). Ch. 4, 'The Pythagorean Principles', is an analysis of the Pythagorean orientation.

MOURELATOS, A., 'Astronomy and Kinematics in Plato's Project of Rationalist Explanation', *Stud. Hist. Phil. Sci.* 12 (1981), 1–32.

PHILIP, J. A., *Pythagoras and Early Pythagoreanism* (Toronto: University of Toronto Press, 1966).

PTOLEMY, C., *The Almagest*, trans. C. Taliaferro, in *Great Books of the Western World*, xvi (Chicago: Encyclopaedia Britannica, 1952).

VLASTOS, G., *Plato's Universe* (Oxford: Clarendon Press, 1975).

3. The Ideal of Deductive Systematization

DIJKSTERHUIS, E. J., *Archimedes*, trans. C. Dikshoorn (Copenhagen: E. Munksgaard, 1956).

Euclid, *Elements*, ed. T. L. Heath, 3 vols. (New York: Dover Publications, 1926).

The Works of Archimedes with The Method of Archimedes, ed. T. L. Heath (New York: Dover Publications, n.d., repr. of 1912 Cambridge University Press publication).

5. Affirmation and Development of Aristotle's Method in the Medieval Period

General Works on the Medieval Period

CLAGETT, M., *The Science of Mechanics in the Middle Ages* (Madison, Wis.: University of Wisconsin Press), 1959.

CROMBIE, A. C., *Robert Grosseteste and the Origins of Experimental Science (1100–1700)* (Oxford: Clarendon Press, 1962); contains an extensive bibliography.

GRANT, E. (ed.), *A Source Book in Medieval Science* (Cambridge, Mass.: Harvard University Press, 1974).

KRETZMANN, N., *et al.* (eds.), *The Cambridge History of Later Medieval Philosophy* (Cambridge: Cambridge University Press, 1982), chs. 6 and 7.

MOODY, E. A., 'Empiricism and Metaphysics in Medieval Philosophy', *Phil. Rev.* 67 (1958), 145–63.

SHAPIRO, H. (ed.), *Medieval Philosophy, Selected Readings, from Augustine to Buridan* (New York: The Modern Library, 1964).

SHARP D. E., *Franciscan Philosophy at Oxford in the Thirteenth Century* (New York: Russell & Russell, 1964).

THORNDIKE, L., *A History of Magic and Experimental Science*, ii (New York: Macmillan, 1923).

WALLACE, W. A., *Causality and Scientific Explanation*, i (Ann Arbor, Mich.: University of Michigan Press, 1972).

WEINBERG, J. R., *A Short History of Medieval Philosophy* (Princeton, NJ: Princeton University Press, 1964).

—— 'Historical Remarks on Some Medieval Views of Induction', in J. R. Weinberg, *Abstraction, Relation, and Induction* (Madison, Wis.: University of Wisconsin Press, 1965), 121–53.

Robert Grosseteste

CROMBIE, A. C., 'Grosseteste's Position in the History of Science', in D. A. Callus (ed.), *Robert Grosseteste* (Oxford: Clarendon Press, 1955).

—— 'Quantification in Medieval Physics', *Isis*, 52 (1961), 143–60.

DALES, R. C., 'Robert Grosseteste's Scientific Works', *Isis*, 52 (1961), 381–402.

McEVOY, J., *The Philosophy of Robert Grosseteste* (Oxford: Clarendon Press, 1982).

SERENE, E., 'Robert Grosseteste on Induction and Demonstrative Science', *Synthèse*, 40 (1979), 97–115.

Roger Bacon

EASTON, S. C., *Roger Bacon and His Search for a Universal Science* (New York: Columbia University Press, 1952).

LINDBERG, D., 'On the Applicability of Mathematics to Nature: Roger Bacon and His Predecessors', *Brit. J. Hist. Sci.* 15 (1982), 3–26.

The Opus Majus, trans. R. B. Burke (New York: Russell & Russell, 1962).

STEELE, R., 'Roger Bacon and the State of Science in the Thirteenth Century', in C. Singer (ed.), *Studies in the History and Method of Science* (Oxford: Clarendon Press, 1921), ii. 121–50.

John Duns Scotus

BOLER, J. F., *Charles Peirce and Scholastic Realism* (Seattle: University of Washington Press, 1963), 37–62.

Duns Scotus: Philosophical Writings, ed. and trans. A. B. Wolter (Edinburgh: Nelson, 1962).

HARRIS, C. R. S., *Duns Scotus* (1927), 2 vols. (New York: Humanities Press, 1959).

William of Ockham

BOEHNER, P., *Collected Articles on Ockham*, ed. E. M. Buytaert (St Bonaventure, NY: Franciscan Institute Publications, 1958).

MAURER, A., 'Method in Ockham's Nominalism', *Monist*, 61 (1978), 426–43.

MOODY, E. A., 'Ockham, Buridan, and Nicolaus of Autrecourt', *Franciscan Stud.* 7 (1947), 115–46.

—— *The Logic of William of Ockham* (New York: Russell & Russell, 1965).

Ockham: Philosophical Writings, ed. with an introduction by P. Boehner (Edinburgh: Nelson, 1962); contains a bibliography of Ockham's works.

Ockham: Studies and Selections, ed. with an introduction by S. C. Tornay (La Salle, Ill.: Open Court Publishing Co., 1938).

SHAPIRO, H., *Motion, Time and Place According to William Ockham* (St Bonaventure, NY: Franciscan Institute Publications, 1957).

TWEEDALE, M., 'Abailard and Ockham's Contrasting Defenses of Nominalism', *Theoria*, 46 (1980), 106–22.

Nicolaus of Autrecourt

'First and Second Letters to Bernard of Arezzo', in H. Shapiro (ed.), *Medieval Philosophy*, Selected Readings, from Augustine to Buridan

(New York: The Modern Library, 1964), 510–27.

WEINBERG, J. R., *Nicolaus of Autrecourt: A Study in Fourteenth-Century Thought* (Princeton, NJ: Princeton University Press, 1948).

6. The Debate Over Saving the Appearances

BAIGRIE, B., 'Kepler's Laws of Planetary Motion, Before and After Newton's *Principia*', *Stud. Hist. Phil. Sci.* 18 (1987), 177–208.

DRAKE, S., 'Hipparchus-Geminus-Galileo', *Stud. Hist. Phil. Sci.* 20 (1989), 47–56.

DUHEM, P., *To Save the Phenomena*, trans. E. Doland and C. Maschler (Chicago: University of Chicago Press, 1969).

FIELD, J. V., *Kepler's Geometrical Cosmology* (Chicago: University of Chicago Press, 1988).

JARDINE, N., *The Birth of History and Philosophy of Science: Kepler's Defence of Tycho Against Ursus, with Essays on its Provenance and Significance* (Cambridge: Cambridge University Press, 1984).

KEPLER, J., *Mysterium Cosmographicum*, trans. A. M. Duncan (New York: Abaris Books, 1981).

KOYRÉ, A., *La Révolution astronomique* (Paris: Hermann, 1961).

KUHN, T. S., *The Copernican Revolution* (New York: Random House, 1957).

O'NEIL, W. M., *Fact and Theory*, Pt. 2 (Sydney: Sydney University Press, 1969).

Ptolemy, Copernicus, Kepler, in *Great Books of the Western World*, xvi (Chicago: Encyclopaedia Britannica, 1952); contains:

Ptolemy, *The Almagest*, trans. R. C. Taliaferro.

Copernicus, *On the Revolutions of the Heavenly Spheres*, trans. C. G. Wallis.

Kepler, *Epitome of Copernican Astronomy*, bk. 5, trans. C. G. Wallis.

Three Copernican Treatises, 2nd edn., trans. E. Rosen (New York: Dover Publications, 1959); contains:

Copernicus, *Commentariolis*.

Copernicus, *Letter Against Werner*.

Rheticus, *Narratio Prima*.

Annotated Copernicus Bibliography (1939–58), compiled by Rosen.

WESTMAN, R. S. (ed.), *The Copernican Achievement* (Berkeley, Calif.: University of California Press, 1975).

7. The Seventeenth-Century Attack on Aristotelian Philosophy

I. Galileo

Works By Galileo

The Assayer, trans. S. Drake, in *The Controversy on the Comets of 1618*, trans. S. Drake and C. D. O'Malley (Philadelphia: University of Pennsylvania Press, 1960), 151–336.

Dialogue Concerning the Two Chief World Systems (1632), trans. S. Drake (Berkeley, Calif.: University of California Press, 1953).

Discoveries and Opinions of Galileo, trans. S. Drake (Garden City, NY: Doubleday Anchor Books, 1957); includes *The Starry Messenger* (1610); *Letters on Sunspots* (1613); *Letter to the Grand Duchess Christina* (1615); and a portion of the *Assayer* (1623).

Two New Sciences (1638), trans. S. Drake (Madison, Wis.: University of Wisconsin Press, 1974).

Works about Galileo

BUTTS, R. E., and PITT, J. C. (eds.), *New Perspectives on Galileo* (Dordrecht: Reidel, 1978).

DE SANTILLANA, G., *The Crime of Galileo* (Chicago: University of Chicago Press, 1963).

DRAKE, S., *Galileo Studies* (Ann Arbor, Mich.: University of Michigan Press, 1970).

FEHÉR, M., 'Galileo and the Demonstrative Ideal of Science', *Stud. Hist. Phil. Sci.* 13 (1982), 87–110.

FINOCCHIARO, M., *Galileo and the Art of Reasoning* (Dordrecht: Reidel, 1980).

GEYMONAT, L., *Galileo Galilei*, trans. S. Drake (New York: McGraw-Hill, 1965).

GOOSENS, W., 'Galileo's Response to the Tower Argument', *Stud. Hist. Phil. Sci.* 11 (1980), 215–27.

KOERTGE, N., 'Galileo and the Problem of Accidents', *J. Hist. Ideas* 38 (1977), 389–408.

KOYRÉ, A., 'Galileo and Plato', *J. Hist. Ideas* 4 (1943), 400–28.

—— 'Galileo and the Scientific Revolution of the Seventeenth Century', *Phil Rev.* 52 (1943), 333–48.

—— 'An Experiment in Measurement', *Proc. Am. Phil. Soc.* 97 (1953), 222–37.

MCMULLIN, E., (ed.), *Galileo, Man of Science* (New York: Basic Books, 1967).

—— 'Galilean Idealization', *Stud. Hist. Phil. Sci.* 16 (1985), 247–73.

MERTZ, D., 'The Concept of Structure in Galileo', *Stud. Hist. Phil. Sci.* 13 (1982), 111–31.

REDONDI, P., *Galileo Heretic* (Princeton, NJ: Princeton University Press, 1987).

SHAPERE, D., *Galileo* (Chicago: University of Chicago Press, 1974).

SHEA, W., *Galileo's Intellectual Revolution* (New York: Science History, 1972).

WALLACE, W. A., *Galileo and His Sources* (Princeton, NJ: Princeton University Press, 1984).

II. Francis Bacon

Works by Francis Bacon

The Works of Francis Bacon, 14 vols., ed. J. Spedding, R. L. Ellis, and D. D. Heath (New York: Hurd and Houghton, 1869).

Works about Francis Bacon

ANDERSON, F. H., *The Philosophy of Francis Bacon* (Chicago: University of Chicago Press, 1948).

BROAD, C. D., *The Philosophy of Francis Bacon* (Cambridge: Cambridge University Press, 1926).

COHEN, L. J., 'Some Historical Remarks on the Baconian Conception of Probability', *J. Hist. Ideas* 41 (1980), 219–31.

DUCASSE, C. J., 'Francis Bacon's Philosophy of Science', in R. M. Blake, C. J. Ducasse, and E. H. Madden (eds.), *Theories of Scientific Method: The Renaissance Through the Nineteenth Century* (Seattle: University of Washington Press, 1960).

FARRINGTON, B., *Francis Bacon: Philosopher of Industrial Science* (New York: Schuman, 1949).

—— *The Philosophy of Francis Bacon: An Essay on its Development from 1603 to 1609 with New Translations of Fundamental Texts* (Liverpool: Liverpool University Press, 1964).

PRIMACK, M., 'Outline of a Reinterpretation of Francis Bacon's Philosophy', *J. Hist. Phil.* 5 (1967), 123–32.

ROSSI, P., *Francis Bacon: From Magic to Science*, trans. S. Rabinovitch (London: Routledge & Kegan Paul, 1968).

URBACH, P., *Francis Bacon's Philosophy of Science* (La Salle, Ill.: Open Court, 1987).

III. Descartes

Works by Descartes

Descartes: Philosophical Letters, trans. and ed. A. Kenny (Oxford: Clarendon Press, 1970).

Descartes: Philosophical Writings, ed. and trans. G. E. M. Anscombe and P. T. Geach (Edinburgh: Nelson, 1954).

Œuvres de Descartes, ed. by C. Adam and P. Tannery (Paris: Leopold Cerf, 1897–1913).

The Philosophical Works of Descartes, trans. E. S. Haldane and G. R. T. Ross, 2 vols. (New York: Dover Publications, 1955).

Principles of Philosophy, trans. V. R. and R. P. Miller (Dordrecht: Reidel, 1983).

Works about Descartes

AYER, A. J., 'Cogito ergo sum', *Analysis*, 14 (1953), 27–31.

BECK, L. J., *The Method of Descartes: A Study of the Regulae* (Oxford: Clarendon Press, 1952).

BECK, L. J., *The Metaphysics of Descartes: A Study of the Meditations* (Oxford: Clarendon Press, 1965).

BLAKE, R. M., 'The Role of Experience in Descartes' Theory of Method', *Phil. Rev.* 38 (1929), 125–43, 201–18. Repr. in R. M. Blake, C. J. Ducasse, and E. H. Madden, *Theories of Scientific Method: The Renaissance Through the Nineteenth Century* (Seattle: University of Washington Press, 1960).

BROUGHTON, J., 'Skepticism and the Cartesian Circle', *Can. J. Phil.* 14 (1984), 593–615.

BUCHDAHL, G., 'The Relevance of Descartes's Philosophy for Modern Philosophy of Science', *Brit. J. Hist. Sci.* 1 (1963), 227–49.

BUTLER, R. J. (ed.), *Cartesian Studies* (Oxford: Blackwell, 1972).

CHAPPELL, V. (ed.), *Twenty-Five Years of Descartes Scholarship, 1960–1984: A Bibliography* (New York: Garland, 1987); updates the *Bibliographia Cartesiana*.

CLARKE, D., *Descartes' Philosophy of Science* (Manchester: Manchester University Press, 1982).

CURLEY, E. M., *Descartes Against the Skeptics* (Cambridge, Mass: Harvard University Press, 1978).

DONEY, W. (ed.), *Descartes: A Collection of Critical Essays* (Garden City, NY: Doubleday, 1967).

GAUKROGER, S. (ed.), *Descartes: Philosophy, Mathematics and Physics* (Totowa, NJ: Barnes & Noble, 1980).

—— *Cartesian Logic* (Oxford: Clarendon Press, 1989).

HATFIELD, G., 'Force (God) in Descartes' Physics', *Stud. Hist. Phil. Sci.* 10 (1979), 113–40.

HOOKER, M. (ed.), *Descartes: Critical and Interpretive Essays* (Baltimore: The Johns Hopkins Press, 1978).

OSLER, M. J., 'Eternal Truths and the Laws of Nature: The Theological Foundations of Descartes' Philosophy of Nature', *J. Hist. Ideas* 46 (1985), 349–62.

PASSMORE, J. A., 'William Harvey and the Philosophy of Science', *Australasian J. Phil.* 36 (1958), 85–94.

RADNER, D., 'Is There a Problem of Cartesian Interaction?', *J. Hist. Phil.* 23 (1985), 35–49.

SEBBA, G., *Bibliographia Cartesiana: A Critical Guide to the Descartes Literature* (1800–1960) (The Hague: Martinus Nijhoff, 1964).

SESONSKE, A., and FLEMING, N. (eds.), *Meta-Meditations: Studies in Descartes* (Belmont, Calif.: Wadsworth, 1965).

SMITH, W. K., *New Studies in the Philosophy of Descartes* (New York: Russell & Russell, 1966).

SUPPES, P., 'Descartes and the Problem of Action at a Distance', *J. Hist. Ideas* 15 (1954), 146–52.

WILLIAMS, B., *Descartes: The Project of Pure Enquiry* (Atlantic Highlands, NJ: Humanities Press, 1978).

8. Newton's Axiomatic Method

Works by Newton

Isaac Newton's Papers and Letters on Natural Philosophy, ed. I. B. Cohen (Cambridge, Mass: Harvard University Press, 1958).

Newton's Mathematical Principles of Natural Philosophy and His System of the World, trans. A. Motte (1729), rev. F. Cajori, 2 vols. (Berkeley, Calif.: University of California Press, 1962).

Opticks, 4th edn. (1730) (New York: Dover Publications, 1952).

Unpublished Scientific Papers of Isaac Newton, ed. and trans. A. R. and M. B. Hall (Cambridge: Cambridge University Press, 1962).

Works about Newton

BECHLER, Z., *Contemporary Newtonian Research* (Dordrecht: Reidel, 1982); essays by I. B. Cohen, R. S. Westfall, J. E. McGuire, *et al.*

BLAKE, R. M., 'Isaac Newton and the Hypothetico-Deductive Method' in R. M. Blake, C. J. Ducasse, and E. H. Madden, *Theories of Scientific Method: The Renaissance Through the Nineteenth Century* (Seattle: University of Washington Press, 1960), 119–43.

BOAS, M., and HALL, A. R., 'Newton's "Mechanical Principles"', *J. Hist. Ideas* 20 (1959), 167–78.

BRICKER, P., and HUGHES, R. I., *Philosophical Perspectives in Newtonian Science* (Cambridge, Mass.: MIT Press, 1990).

BUCHDAHL, G., 'Science and Logic: Some Thoughts on Newton's Second Law of Motion in Classical Mechanics', *Brit. J. Phil. Sci.* 2 (1951–2), 217–35.

BUTTS, R. E., and DAVIS, J. W. (eds.), *The Methodological Heritage of*

Newton (Toronto: University of Toronto Press, 1970); a collection of critical essays.

COHEN, I. B., *Franklin and Newton* (Cambridge, Mass.: Harvard University Press, 1966).

—— *The Newtonian Revolution* (Cambridge: Cambridge University Press, 1980).

—— 'Newton's Third Law and Universal Gravity', *J. Hist. Ideas* 48 (1987), 571–93.

FAUVEL, J., *et al.* (eds.), *Let Newton Be* (Oxford: Oxford University Press, 1988); essays on Newton's achievements in mathematics, science, and theology.

FEHÉR, M., 'The Method of Analysis-Synthesis and the Structure of Causal Explanation in Newton', *Int. Stud. Phil. Sci.* 1 (1986), 60–84.

HALL, A. R., *Philosophers At War: The Quarrel Between Newton and Leibniz* (Cambridge: Cambridge University Press, 1980).

KOYRÉ, A., *Newtonian Studies* (Cambridge, Mass.: Harvard University Press, 1965).

LAYMON, R., 'Newton's Bucket Experiment', *J. Hist. Phil.* 16 (1978), 399–413.

McMULLIN, E., *Newton on Matter and Activity* (Notre Dame, Ind.: University of Notre Dame Press, 1978).

MANUEL, F., *A Portrait of Isaac Newton* (Cambridge, Mass.: Harvard University Press, 1968).

The Texas Quarterly, 10 (Autumn 1967) (Austin, Tex.: University of Texas Press); contains articles on Newton by I. B. Cohen, A. R. and M. B. Hall, J. Herivel, R. S. Westfall, *et al.*

WESTFALL, R. S., *Force in Newton's Physics* (London: MacDonald, 1971).

—— *Never at Rest: A Biography of Isaac Newton* (Cambridge: Cambridge University Press, 1980).

9. Analyses of the Implications of the New Science for a Theory of Scientific Method

I. The Cognitive Status of Scientific Laws

General

BUCHDAHL, G., *Metaphysics and the Philosophy of Science* (Oxford: Blackwell, 1969).

WALLACE, W. A., *Causality and Scientific Explanation*, ii (Ann Arbor, Mich.: University of Michigan Press, 1972).

Works by Locke

An Essay Concerning Human Understanding, 1st edn. (1690), 2 vols. (New York: Dover Publications, 1959).

Works of John Locke, 10th edn., 10 vols. (London: J. Johnson, 1801).

Works about Locke

AARON, R. I., *John Locke*, 2nd edn. (Oxford: Clarendon Press, 1955).

GIBSON, J., *Locke's Theory of Knowledge* (Cambridge: Cambridge University Press, 1917).

HEIMANN, P. M., and McGUIRE, J. E., 'Newtonian Forces and Lockean Powers: Concepts of Matter in Eighteenth-Century Thought', *Hist. Stud. Phys. Sci.* 3 (1971), 233–306.

LAUDAN, L., 'The Nature and Sources of Locke's Views on Hypotheses', *J. Hist. Ideas* 28 (1967), 211–23.

LENNON, J. M., 'Locke's Atomism', *Phil. Res. Archives* 9 (1983), 1–28.

MANDELBAUM, M., *Philosophy, Science and Sense Perception: Historical and Critical Studies* (Baltimore: The Johns Hopkins Press, 1964), ch. 1.

MATTERN, R. M., 'Locke on Active Power and the Obscure Idea of Active Power from Bodies', *Stud. Hist. Phil. Sci.* 11 (1980), 39–77.

MARTIN, C. B., and ARMSTRONG, D. M., *Locke and Berkeley* (Garden City, NY: Doubleday & Co., 1968).

O'CONNOR, D. J., *John Locke* (New York: Dover Publications, 1967).

YOLTON, J. W., *John Locke and the Way of Ideas* (Oxford: Clarendon Press, 1956).

YOST, R. M., 'Locke's Rejection of Hypotheses About Sub-Microscopic Events', *J. Hist. Ideas* 12 (1951), 111–30.

Works by Leibniz

Die philosophischen Schriften von G. W. Leibniz, 7 vols., ed. C. I. Gerhardt (Berlin: Weidmann, 1875–90).

Leibniz: Philosophical Papers and Letters, trans. and ed. L. E. Loemker (Dordrecht: D. Reidel Publishing Co., 1969); contains an extensive bibliography.

Leibniz Selections, ed. P. Wiener (New York: Charles Scribner's Sons, 1951).

Works about Leibniz

AITON, E. J., *Leibniz: A Biography* (Bristol: Adam Hilger, 1985).

FRANKFURT, H. G. (ed.), *Leibniz: A Collection of Critical Essays* (Garden City, NY: Doubleday, 1972).

GALE, G., 'The Concept of "Force" and Its Role in the Genesis of Leibniz' Dynamical Viewpoint', *J. Hist. Phil.* 26 (1988), 45–67.

OKRUHLIK, K., and BROWN, J. R. (eds.), *The Natural Philosophy of Leibniz* (Dordrecht: Reidel, 1985).

RESCHER, H., *The Philosophy of Leibniz* (Englewood Cliffs, NJ: Prentice Hall, 1967).

RUSSELL, B., *A Critical Exposition of the Philosophy of Leibniz*, 2nd edn. (London: George Allen & Unwin, 1937).

WILSON, C., 'Leibniz and Atomism', *Stud. Hist. Phil. Sci.* 13 (1982), 175–200.

WINTERBOURNE, A. T., 'On the Metaphysics of Leibnizian Space and Time', *Stud. Hist. Phil. Sci.* 13 (1982), 201–14.

Works by Hume

An Enquiry Concerning Human Understanding (1748) (Chicago: Open Court Publishing Co., 1927).

A Treatise of Human Nature (1739–40), ed. L. A. Selby-Bigge (Oxford: Clarendon Press, 1965).

Hume's Philosophical Works, ed. T. H. Green and T. H. Grose, 4 vols. (London: Longmans, 1874–5).

Works about Hume

BEAUCHAMP, T., and ROSENBERG, A., *Hume and the Problem of Causation* (Oxford: Oxford University Press, 1981).

BROUGHTON, J., 'Hume's Ideas About Necessary Connection', *Hume Stud.* 13 (1987), 217–44.

COSTA, M., 'Hume and Causal Inference', *Hume Stud.* 12 (1986), 141–59.

FLEW, A., *Hume's Philosophy of Belief* (New York: Humanities Press, 1961).

Human Understanding: Studies in the Philosophy of David Hume, ed. A. Sesonske and N. Fleming (Belmont, Calif.: Wadsworth Publishing Company, 1965).

Hume, ed. V. C. Chappell (Garden City, NY: Doubleday & Co., 1966).

JESSOP, T. E., *Bibliography of David Hume and of Scottish Philosophy from Francis Hutcheson to Lord Balfour* (1938) (New York: Russell & Russell, 1966).

MOORE, G. E., 'Hume's Philosophy', in *Philosophical Studies* (New York: Harcourt, Brace & Co., 1922). Repr. in *Readings in Philosophical Analysis*, ed. H. Feigl and W. Sellars (New York: Appleton-Century-Crofts, 1949), 351–63.

PRICE, H. H., *Hume's Theory of the External World* (Oxford: Clarendon Press, 1940).

SMITH, N. K., *The Philosophy of David Hume* (London: Macmillan, 1941).

WILL, F. L., 'Will the Future Be Like the Past?', *Mind*, 56 (1947), 332–47.

YOLTON, J. W., 'The Concept of Experience in Locke and Hume', *J. Hist. Phil.* i (1963), 53–72.

Works by Kant

Immanuel Kant's 'Critique of Pure Reason', trans. F. M. Muller, 2nd edn. (1896) (New York: Macmillan, 1934).

Kant's Gesammelte Schriften, ed. under the supervision of the Berlin Academy of Sciences, 23 vols. (Berlin: Georg Reimer, 1902–55).

Kant's Kritik of Judgement, trans. J. H. Bernard (London: Macmillan, 1892).

Metaphysical Foundations of Natural Science, trans. J. Ellington (Indianapolis: Bobbs-Merrill, 1970).

Works about Kant

BECK, L. W., *Studies in the Philosophy of Kant* (Indianapolis: Bobbs-Merrill, 1965).

BENNETT, J. F., *Kant's Analytic* (Cambridge: Cambridge University Press, 1966).

BIRD, G., *Kant's Theory of Knowledge* (New York: Humanities Press, 1962).

BRITTAN, G. G., *Kant's Theory of Science* (Princeton, NJ: Princeton University Press, 1978).

BUCHDAHL, G., 'Causality, Causal Laws and Scientific Theory in the Philosophy of Kant', *Brit. J. Phil. Sci.* 16 (1965–6), 187–208.

—— 'The Kantian "Dynamic of Reason", with Special Reference to the Place of Causality in Kant's System', in L. W. Beck (ed.), *Kant Studies Today* (La Salle, Ill.: Open Court, 1969), 341–71.

—— 'The Conception of Lawlikeness in Kant's Philosophy of Science', *Synthèse*, 23 (1971), 24–46.

BUTTS, R. E., 'On Buchdahl's and Palter's Papers', *Synthèse*, 23 (1971), 63–74.

—— (ed.), *Kant's Philosophy of Physical Science* (Dordrecht: Reidel, 1986).

GRAM, M. S. (ed.), *Kant: Disputed Questions* (Chicago: Quadrangle Books, 1967).

GUYER, P., 'Kant's Conception of Empirical Law', *Arist. Soc. Supp.* 64 (1990), 221–42.

KITCHER, P., 'Kant's Philosophy of Science', *Midwest Stud. Phil.* 8 (1983), 387–407; repr. in A. Wood (ed.), *Self and Nature in Kant's Philosophy* (Ithaca, NY: Cornell University Press, 1984).

KÖRNER, S., *Kant* (Harmondsworth: Penguin, 1960).

PALTER, R., 'Absolute Space and Absolute Motion in Kant's Critical Philosophy', *Synthèse*, 23 (1971), 47–62.

RESCHER, N., 'On the Status of "Things in Themselves" in Kant's Philosophy', *Synthèse*, 47 (1981), 289–99.

SMITH, N. K., *A Commentary to Kant's 'Critique of Pure Reason'*, 2nd edn. (1923) (New York: Humanities Press, 1962).

STRAWSON, P., *The Bounds of Sense: An Essay on Kant's 'Critique of Pure Reason'* (London: Methuen, 1966).

WALKER, R. C. S., *Kant* (London: Routledge & Kegan Paul, 1978).

WHITNEY, G. T., and BOWERS, D. F. (eds.), *The Heritage of Kant* (Princeton, NJ: Princeton University Press, 1939).

WOLFF, R. P. (ed.), *Kant* (Garden City, NY: Doubleday & Co., 1967).

II. *Theories of Scientific Procedure*

A Preliminary Discourse on the Study of Natural Philosophy (London: Longman, Rees, Orme, Brown & Green, and John Taylor, 1830).

Familiar Lectures on Scientific Subjects (New York: George Routledge & Sons, 1871).

Outlines of Astronomy, 2 vols. (New York: P. F. Collier & Son, 1902).

Works about Herschel

DUCASSE, C. J., 'John F. W. Herschel's Methods of Experimental Inquiry' in R. M. Blake, C. J. Ducasse, and E. H. Madden (eds.), *Theories of Scientific Method: The Renaissance Through the Nineteenth Century* (Seattle: University of Washington Press, 1960), 153–82.

CANNON, W. F., 'John Herschel and the Idea of Science', *J. Hist. Ideas* 22 (1961), 215–39.

Works by Whewell

Astronomy and General Physics Considered with Reference to Natural Theology (Philadelphia: Carey, Lea, & Blanchard, 1836).

The Historical and Philosophical Works of William Whewell, ed. G. Buchdahl and L. Laudan (London: Frank Cass 1967–   ).

History of the Inductive Sciences (1837), 3 vols. (New York: D. Appleton & Co., 1859).

The Philosophy of the Inductive Sciences, 2nd edn., 2 vols. (London: J. W. Parker, 1847), 3rd edn. expanded into 3 parts: *The History of Scientific Ideas*, 2 vols. (London: J. W. Parker & Son, 1858); *Novum Organon Renovatum*, 3rd edn. (London: J. W. Parker & Son, 1858); and *On the Philosophy of Discovery* (London: J. W. Parker & Son, 1860).

William Whewell's Theory of Scientific Method, ed. R. E. Butts (Pittsburgh: University of Pittsburgh Press, 1968); contains selections from Whewell's writings, a bibliography of works by and about Whewell, and an introductory essay by Butts.

Works about Whewell

ACHINSTEIN, P., 'Hypotheses, Probability, and Waves', *Brit. J. Phil. Sci.* 41 (1990), 73–102. An evaluation of the competing views of Whewell and Mill.

BUTTS, R. E., 'Necessary Truth in Whewell's Philosophy of Science', *Am. Phil. Quart.* 2 (1965), 161–81.

—— 'On Walsh's Reading of Whewell's View of Necessity', *Phil. Sci.* 32 (1965), 175–81.

—— 'Whewell's Logic of Induction', in R. N. Giere and R. S. Westfall (eds.), *Foundations of Scientific Method: The Nineteenth Century* (Bloomington, Ind.: Indiana University Press, 1973), 53–85.

DUCASSE, C. J., 'Whewell's Philosophy of Scientific Discovery', *Phil. Rev.* 60 (1951), 56–69; 213–34; repr. in R. M. Blake, C. J. Ducasse, and E. H. Madden (eds.), *Theories of Scientific Method: The Renaissance Through the Nineteenth Century* (Seattle: University of Washington Press, 1960), ch. 9.

FISCH, M., 'Necessary and Contingent Truth in William Whewell's Antithetical Theory of Knowledge', *Stud. Hist. Phil. Sci.* 16 (1985), 275–314.

HEATHCOTE, A. W., 'William Whewell's Philosophy of Science', *Brit. J. Phil. Sci.* 4 (1953–4), 302–14.

METCALFE, J., 'Whewell's Developmental Psychologism: A Victorian Account of Scientific Progress', *Stud. Hist. Phil. Sci.* 22 (1991), 117–39.

STRONG, E. W., 'William Whewell and John Stuart Mill: Their Controversy about Scientific Knowledge', *J. Hist. Ideas* 16 (1955), 209–31.

WALSH, H. T., 'Whewell and Mill on Induction', *Phil. Sci.* 29 (1962), 279–84.

—— 'Whewell on Necessity', *Phil. Sci.* 29 (1962), 139–45.

Works by Meyerson

De l'explication dans les sciences (Paris: Payot, 1927).

Du cheminement de la pensée, 3 vols. (Paris: F. Alcan, 1931).

Identity and Reality (1908), trans. K. Loewenberg (New York: Dover Publications, 1962).

La Déduction rélativiste (Paris: Payot, 1925).

Réel et determinisme dans la physique (Paris: Hermann, 1933).

Works about Meyerson

BOAS, G. A., *A Critical Analysis of the Philosophy of Émile Meyerson* (Baltimore: The Johns Hopkins Press, 1930).

HILLMAN, O. N., 'Émile Meyerson on Scientific Explanation', *Phil. Sci.* 5 (1938), 73–80.

KELLY, T. R., *Explanation and Reality in the Philosophy of Émile Meyerson* (Princeton, NJ: Princeton University Press, 1937).

LaLumia, J., *The Ways of Reason: A Critical Study of the Ideas of Émile Meyerson* (New York: Humanities Press, 1966).

Zahar, E., 'Meyerson's "Relativistic Deduction": Einstein Versus Hegel', *Brit. J. Phil. Sci.* 38 (1987), 93–116.

III. Structure of Scientific Theories

Works by Duhem

The Aim and Structure of Physical Theory, 2nd edn. (1914), trans. P. P. Wiener (New York: Atheneum, 1962).

Études sur Léonard de Vinci, 3 vols. (Paris: A. Hermann, 1906–13).

Le Système du monde: Histoire des doctrines cosmologiques de Platon à Copernic, 5 vols. (Paris: A. Hermann et fils, 1913–17); reissued, 6 vols. (1954).

To Save the Phenomena, trans. E. Doland and C. Maschler, (Chicago: University of Chicago Press, 1969).

Works about Duhem

Ariew, R., 'The Duhem Thesis', *Brit. J. Phil. Sci.* 35 (1984), 313–25.

—— and Barker, P. (eds.), 'Pierre Duhem: Historian and Philosopher of Science', *Synthèse*, 83 (1990), 179–453; essays by A. Brenner, A. Goddu, R. Maiocchi, R. S. Westman, *et al.*

Harding, S. (ed.), *Can Theories Be Refuted? Essays on the Duhem–Quine Thesis* (Dordrecht: Reidel, 1976); essays by A. Grünbaum, M. B. Hesse, L. Laudan, *et al.*

Krips, H., 'Epistemological Holism: Duhem or Quine?' *Stud. Hist. Phil. Sci.* 13 (1982), 251–64.

Tuana, N., 'Quinn on Duhem: An Emendation', *Phil. Sci.* 45 (1978), 456–62; rejoinder by P. Quinn, ibid. 463–5.

Vuillemin, J., 'On Duhem's and Quine's Theses', in L. Hahn (ed.), *The Philosophy of W. V. Quine* (La Salle, Ill.: Open Court, 1986).

Works by Campbell

Foundations of Science, formerly *Physics: The Elements* (1919) (New York: Dover Publications, 1957).

What is Science? (1921) (New York: Dover Publications, 1952).

Works about Campbell

Hempel, C. G. *Aspects of Scientific Explanation and Other Essays in the Philosophy of Science* (New York: Free Press, 1965), 206–10, 442–7.

Hesse, M. B., *Models and Analogies in Science* (New York: Sheed & Ward, 1963).

SCHLESINGER, G., *Method in the Physical Sciences* (New York: Humanities Press, 1963), ch. 3, sect. 5.

Works by Hesse

'An Inductive Logic of Theories', in M. Radner and S. Winokur (eds.), *Minnesota Studies in the Philosophy of Science*, iv (Minneapolis: University of Minnesota Press, 1970), 164–80.

'Analogy and Confirmation Theory', *Phil. Sci.* 31 (1964), 319–27.

'Consilience of Inductions', in I. Lakatos (ed.), *The Problem of Inductive Logic* (Amsterdam: North Holland, 1968), 232–46, 254–7.

Forces and Fields (London: Nelson, 1961).

'Is There an Independent Observation Language?', in R. Colodny (ed.), *The Nature and Function of Scientific Theories* (Pittsburgh: University of Pittsburgh Press, 1970), 35–77.

Models and Analogies in Science (Notre Dame, Ind.: University of Notre Dame Press, 1966).

'Models in Physics', *Brit. J. Phil. Sci.* 4 (1953–4), 198–214.

'Positivism and the Logic of Scientific Theories', in P. Achinstein and S. Barker (eds.), *The Legacy of Logical Positivism* (Baltimore: The Johns Hopkins Press, 1969), 85–114.

Revolutions and Reconstructions in the Philosophy of Science (Bloomington, Ind.: Indiana University Press, 1980).

Science and the Human Imagination (London: SCM Press, 1954).

The Structure of Scientific Inference (London: Macmillan, 1974).

'Theories, Dictionaries, and Observation', *Brit. J. Phil. Sci.* 9 (1958–9), 12–28.

'What is the Best Way to Assess Evidential Support for Scientific Theories?', in L. J. Cohen and M. B. Hesse (eds.), *Applications of Inductive Logic* (Oxford: Clarendon Press, 1980).

Works by Harré

The Anticipation of Nature (London: Hutchinson, 1965).

Causal Powers, with E. H. Madden (Oxford: Blackwell, 1975).

'Concepts and Criteria', *Mind*, 73 (1964), 353–63.

The Explanation of Social Behaviour, with Paul Secord (Oxford: Basil Blackwell, 1972).

An Introduction to the Logic of the Sciences (London: Macmillan, 1967).

Matter and Method (London: Macmillan, 1964).

Philosophies of Science (Oxford: Oxford University Press, 1972).

'Powers', *Brit. J. Phil. Sci.* 21 (1970), 81–101.

The Principles of Scientific Thinking (London: Macmillan, 1970).

Theories and Things (London: Newman History and Philosophy of Science Series, 1961).
Varieties of Realism (Oxford: Blackwell, 1986).

Works about Harré

BHASKAR, R. (ed.), *Harré and His Critics* (Cambridge: Blackwell, 1990).
FRANKEL, H., 'Harré on Causation', *Phil. Sci.* 43 (1976), 560–9.
WILSON, F., 'Dispositions Defined: Harré and Madden on Analysing Disposition Concepts', *Phil. Sci.* 52 (1985), 591–607.

10. Inductivism *v.* the Hypothetico-Deductive View of Science

Works by Mill

A System of Logic: Ratiocinative and Inductive, 6th edn. (London: Longmans, Green, 1865).
Works, ed. F. E. L. Priestley, J. M. Robinson, *et al.* (Toronto: University of Toronto Press, 1963–   ).

Works about Mill

ANSCHUTZ, R. P., *The Philosophy of J. S. Mill* (Oxford: Clarendon Press, 1953).
BRADLEY, F. H., *Principles of Logic*, 2nd edn. (Oxford: Oxford University Press, 1928); bk. 2, pt. II, ch. 3 includes a discussion of Mill's view of induction.
DUCASSE, C. J., 'John Stuart Mill's System of Logic', in R. M. Blake, C. J. Ducasse, and E. H. Madden, *Theories of Scientific Method: The Renaissance through the Nineteenth Century* (Seattle: University of Washington Press, 1960), 218–32.
JACOBS, S., 'John Stuart Mill on Induction and Hypothesis', *J. Hist. Phil.* 29 (1991), 69–83.
JEVONS, W. S., 'John Stuart Mill's Philosophy Tested', pt. 2 of *Pure Logic and Other Minor Works* (London: Macmillan, 1890).
LAINE, M., *Bibliography of Works on John Stuart Mill* (Toronto: University of Toronto Press, 1982); selective, with many brief annotations.
LOSEE, J., 'Whewell and Mill on the Relation Between Philosophy of Science and History of Science', *Stud. Hist. Phil. Sci.* 14 (1983), 113–21.
RYAN, A., *The Philosophy of John Stuart Mill* (London: Macmillan, 1970).
SKORUPSKI, J., *John Stuart Mill* (London: Routledge, 1989).

Works by Jevons

The Principles of Science (1877) (New York: Dover Publications, 1958).

11. Mathematical Positivism and Conventionalism

Works by Berkeley

The Works of George Berkeley, Bishop of Cloyne, 9 vols., ed. A. A. Luce and T. E. Jessop (London: Thomas Nelson & Sons, 1948–57).

Works about Berkeley

ASHER, W., 'Berkeley on Absolute Motion', *H. Phil. Quart.* 4 (1987), 447–66.
ATHERTON, M., 'Corpuscles, Mechanism and Essentialism in Berkeley and Locke', *J. Hist. Phil.* 29 (1991), 47–67.
MYHILL, J., 'Berkeley's *De Motu*—An Anticipation of Mach', in *George Berkeley: Lectures Delivered Before the Philosophical Union of the University of California* (Berkeley, Calif.: University of California Press, 1957), 141–57.
NEWTON-SMITH, W. H., 'Berkeley's Philosophy of Science', in J. Foster and H. Robinson (eds.), *Essays on Berkeley: A Tercentennial Celebration* (Oxford: Clarendon Press, 1985).
PITCHER, G., *Berkeley* (London: Routledge & Kegan Paul, 1977).
POPPER, K. R., 'A Note on Berkeley as Precursor of Mach', *Brit. J. Phil. Sci.* 4 (1953–4), 26–36.
SOSA, E. (ed.), *Essays on the Philosophy of George Berkeley* (Dordrecht: Reidel, 1987).
URMSON, J. O., *Berkeley* (Oxford: Oxford University Press, 1982).
WHITROW, G. J., 'Berkeley's Philosophy of Motion', *Brit. J. Phil. Sci.* 4 (1953–4), 37–45.
WINKLER, K., 'Berkeley on Volition, Power, and the Complexity of Causation', *H. Phil. Quart.* 2 (1985), 53–69.

Works by Mach

The Analysis of Sensations (1886), trans. C. M. Williams (New York: Dover Publications, 1959).
History and Root of the Principle of the Conservation of Energy (1872), trans. P. E. Jourdain (Chicago: Open Court Publishing Co., 1910).
Popular Scientific Lectures (1896), trans T. J. McCormack (Chicago: Open Court Publishing Company, 1943).
The Science of Mechanics (1883), trans. T. J. McCormack (La Salle, Ill.: Open Court Publishing Co., 1960).
Space and Geometry (1901–3), trans. T. J. McCormack (Chicago: Open Court Publishing Co., 1906).

Works about Mach

ALEXANDER, P., 'The Philosophy of Science, 1850–1910', in D. J. O'Connor (ed.), *A Critical History of Western Philosophy* (New York: Free Press, 1964), 403–9.

BRADLEY, J., *Mach's Philosophy of Science* (London: Athlone Press, 1971).

BUNGE, M., 'Mach's Critique of Newtonian Mechanics', *Am. J. Phys.* 34 (1966), 585–96.

COHEN, R. S., and SEEGER, R. J. (eds.), 'Ernst Mach, Physicist and Philosopher', *Boston Studies in the Philosophy of Science*, vi (New York: Humanities Press, 1970); contains a bibliography of works by and about Mach.

FEYERABEND, P., 'Mach's Theory of Research and Its Relation to Einstein', *Stud. Hist. Phil. Sci.* 15 (1984), 1–22.

FRANK, P., *Modern Science and Its Philosophy* (New York: George Braziller, 1961), 13–62, 69–95.

LOPARIĆ, Z., 'Problem-Solving and Theory Structure in Mach', *Stud. Hist. Phil. Sci.* 15 (1984), 23–49.

Works by Poincaré

Mathematics and Science: Last Essays, Eng. trans. J. W. Bolduc of *Dernières pensées* (1913) (New York: Dover Publications, 1963).

Science and Hypothesis (1902), trans. G. B. Halsted (New York: Science Press, 1905).

Science and Method (1909), trans. F. Maitland (New York: Dover Publications, 1952).

The Value of Science (1905), trans. G. B. Halsted (New York: Science Press, 1907).

Works about Poincaré

ALEXANDER, P., 'The Philosophy of Science, 1850–1910', in D. J. O'Connor (ed.), *A Critical History of Western Philosophy* (New York: Free Press, 413–17).

KRIPS, H., 'Atomism, Poincaré and Planck', *Stud. Hist. Phil. Sci.* 17 (1986), 43–63.

STUMP, D., 'Henri Poincaré's Philosophy of Science', *Stud. Hist. Phil. Sci.* 20 (1989), 335–63.

Works by Popper

Conjectures and Refutations (New York: Basic Books, 1963).

'The Demarcation Between Science and Metaphysics', in P. A. Schilpp (ed.), *The Philosophy of Rudolf Carnap*, (La Salle, Ill.: Open Court, 1963), 183–226.

'Indeterminism in Quantum Physics and in Classical Physics', *Brit. J. Phil. Sci.* 1 (1950–1), 117–33, 173–95.

The Logic of Scientific Discovery (New York: Basic Books, 1959); 1st edn., *Logik der Forschung* (1934).

'The Nature of Philosophical Problems and their Roots in Science', *Brit. J. Phil. Sci.* 3 (1952–3), 124–56.

'A Note on Natural Laws and So-Called "Contrary-to-Fact Conditionals"', *Mind*, 58 (1949), 62–6.

Objective Knowledge (Oxford: Clarendon Press, 1972).

The Open Society and Its Enemies, 2 vols., 4th edn., rev. (New York: Harper Torchbooks, 1963).

'Philosophy of Science: A Personal Report', in C. A. Mace (ed.), *British Philosophy in the Mid-Century* (London: George Allen & Unwin, 1957), 155–91.

'A Proof of the Impossibility of Inductive Probability', with D. Miller, *Nature*, 302 (1983), 687–8.

'The Propensity Interpretation of Probability', *Brit. J. Phil. Sci.* 10 (1959–60), 25–42.

The Self and Its Brain, with J. Eccles (London: Routledge & Kegan Paul, 1983).

Works about Popper

ACKERMANN, R. J., *The Philosophy of Karl Popper* (Amherst, Mass.: University of Massachusetts Press, 1976).

AGASSI, J., 'To Save Verisimilitude', *Mind*, 90 (1981), 576–9.

BUNGE, M. A. (ed.), *The Critical Approach to Science and Philosophy* (Glencoe, Ill.: Free Press, 1964); a collection of articles, with a bibliography of Popper's publications.

CHIHARA, C. S., and GILLIES, D. A., 'An Interchange on the Popper–Miller Argument', *Phil. Stud.* 54 (1988), 1–8.

DERKSEN, A. A., 'The Alleged Unity of Popper's Philosophy of Science: Falsifiability as Fake Cement', *Phil. Stud.* 48 (1985), 313–36.

FAIN, H., 'Review of *The Logic of Scientific Discovery*', *Phil. Sci.* 28 (1961), 319–24.

NEWTON-SMITH, W. H., *The Rationality of Science* (London: Routledge & Kegan Paul, 1981), ch. 3.

NOLA, R., 'The Status of Popper's Theory of Scientific Method', *Brit. J. Phil. Sci.* 38 (1987), 441–80.

O'HEAR, A., *Karl Popper* (London: Routledge & Kegan Paul, 1980).

SALMON, W., 'Rational Prediction', *Brit. J. Phil. Sci.* 32 (1981), 115–25.

SARKAR, H., *A Theory of Method* (Berkeley, Calif.: University of California Press, 1983), ch. 2.

SCHILPP, P. A. (ed.), *The Philosophy of Karl R. Popper*, 2 vols. (La Salle,

Ill.: Open Court Publishing Co., 1974); contains an 'Intellectual Auto-biography' by Popper, numerous essays on Popper's philosophy, and a bibliography of his writings compiled by T. E. Hansen.

12. Logical Reconstructionist Philosophy of Science

Works in the Logical Reconstructionist Tradition

BRAITHWAITE, R. B., *Scientific Explanation* (Cambridge: Cambridge University Press, 1953).

BRIDGMAN, P. W., *The Logic of Modern Physics* (New York: Macmillan, 1927).

—— *The Nature of Physical Theory* (Princeton, NJ: Princeton University Press, 1936).

—— *Reflections of a Physicist* (New York: Philosophical Library, 1950).

—— *The Way Things Are* (Cambridge, Mass.: Harvard University Press, 1959).

BRODBECK, M. (ed.), *Readings in the Philosophy of the Social Sciences* (Minneapolis: University of Minnesota Press, 1968).

CARNAP, R., 'The Methodological Character of Theoretical Concepts', in H. Feigl and M. Scriven (eds.), *Minnesota Studies in the Philosophy of Science*, i (Minneapolis: University of Minnesota Press, 1956), 38–76.

—— *Logical Foundations of Probability*, 2nd. edn. (Chicago: University of Chicago Press, 1962).

—— *Philosophical Foundations of Physics*, ed. M. Gardner (New York: Basic Books, 1966).

DANTO, A., and MORGENBESSER, S. (eds.), *Philosophy of Science* (New York: Meridian Books, 1960).

FEIGL, H., and BRODBECK, M. (eds.), *Readings in the Philosophy of Science* (New York: Appleton-Century-Crofts, 1953).

FRANK, P., *Philosophy of Science* (Englewood Cliffs, NJ: Prentice-Hall, 1957).

HEMPEL, C., *Aspects of Scientific Explanation* (New York: Free Press, 1965).

—— *Philosophy of Natural Science* (Englewood Cliffs, NJ: Prentice Hall, 1966).

HEMPEL, C., 'Rudolf Carnap: Logical Empiricist', *Synthèse*, 46 (1973), 256–68.

—— 'Turns in the Evolution of the Problem of Induction', *Synthèse*, 46 (1981), 389–404.

HUTTON, E., *The Language of Modern Physics* (London: George Allen & Unwin, 1956).

NAGEL, E., *The Structure of Science* (New York: Harcourt, Brace & World, 1961).

—— 'Theory and Observation', in E. Nagel, S. Bromberger and A. Grünbaum, *Observation and Theory in Science*, ed. M. Mandelbaum (Baltimore: The Johns Hopkins Press, 1971), 15–43.

NEURATH, O., CARNAP, R., and MORRIS, C. (eds.), *Foundations of the Unity of Science*, 2 vols. (formerly, *International Encyclopedia of United Science*, 1938–69) (Chicago: University of Chicago Press, 1969, 1970); includes monographs by R. Carnap, P. Frank, C. Hempel, and others.

PAP, A., *An Introduction to the Philosophy of Science* (Glencoe, Ill.: Free Press, 1962).

RESCHER, N., *Scientific Explanation* (New York: Free Press, 1970).

SMART, J. J. C., *Between Science and Philosophy* (New York: Random House, 1968).

—— *Philosophy and Scientific Realism* (London: Routledge & Kegan Paul, 1963).

Works about the Logical Reconstructionist Tradition

BROWN, H. I., *Perception, Theory and Commitment* (Chicago: University of Chicago Press, 1977).

FEIGL, H., 'Some Major Issues and Developments in the Philosophy of Science of Logical Empiricism', in H. Feigl and M. Scriven (eds.), *Minnesota Studies in the Philosophy of Science*, i (Minneapolis: University of Minnesota Press, 1956), 3–37.

OLDROYD, D., *The Arch of Knowledge* (London: Methuen, 1986), ch. 6.

SCHEFFLER, I., *The Anatomy of Inquiry* (Indianapolis: Bobbs-Merrill, 1963).

SCHILPP, P. (ed.), *The Philosophy of Rudolf Carnap* (La Salle, Ill.: Open Court, 1963); contains an 'Intellectual Autobiography' by Carnap, numerous essays on Carnap's philosophy, and a bibliography of Carnap's writings.

SUPPE, F., 'The Search for Philosophic Understanding of Scientific Theories', in Suppe (ed.), *The Structure of Scientific Theories* (Urbana, Ill.: University of Illinois Press, 1974); contains an extensive bibliography.

13. Orthodoxy under Attack

ACHINSTEIN, P., *Concepts of Science* (Baltimore: The Johns Hopkins Press, 1968).

FEIGL, H., 'Existential Hypotheses', *Phil. Sci.* 17 (1950), 35–62; *Phil. Sci.* 17 also contains criticisms of Feigl's paper by C. Hempel, E. Nagel, and C. W. Churchman, and a rejoinder by Feigl.

—— and MAXWELL, G. (eds.), *Current Issues in the Philosophy of Science* (New York: Holt, Rinehart, and Winston, 1961).

FEYERABEND, P., 'Explanation, Reduction and Empiricism', in H. Feigl and G. Maxwell (eds.), *Minnesota Studies in the Philosophy of Science*, iii (Minneapolis: University of Minnesota Press, 1962), 28–97.

—— 'Problems of Empiricism', in R. Colodny (ed.), *Beyond the Edge of Certainty* (Englewood Cliffs, NJ: Prentice-Hall, 1965).

—— 'How To Be a Good Empiricist—A Plea for Tolerance in Matters Epistemological', in B. Brody (ed.), *Readings in the Philosophy of Science* (Englewood Cliffs, NJ: Prentice-Hall, 1970), 319–42.

—— 'Problems of Empiricism Part II', in R. Colodny (ed.), *The Nature and Function of Scientific Theories* (Pittsburgh: University of Pittsburgh Press, 1970), 275–353.

—— *Against Method* (London: NLB, 1975).

GOODMAN, N., *Fact, Fiction and Forecast*, 2nd edn. (Indianapolis: Bobbs-Merrill, 1965).

GRÜNBAUM, A., 'The Duhemian Argument', *Phil. Sci.* 27 (1960), 75–87.

—— 'The Falsifiability of Theories: Total or Partial? A Contemporary Evaluation of the Duhem-Quine Thesis', in M. Wartofsky (ed.), *Boston Studies in the Philosophy of Science*, i (Dordrecht: D. Reidel, 1963), 178–95.

—— 'Temporally Asymmetric Principles, Parity Between Explanation and Prediction, and Mechanism and Teleology', *Phil. Sci.* 29 (1962), 146–70.

HANSON, N. R., *Patterns of Discovery* (Cambridge: Cambridge University Press, 1958).

MAXWELL, G., 'The Ontological Status of the Theoretical Entities', in H. Feigl and G. Maxwell (eds.), *Minnesota Studies in the Philosophy of Science*, iii (Minneapolis: University of Minnesota Press, 1962).

MICHALOS, A., *The Popper–Carnap Controversy* (The Hague: Martinus Nijhoff, 1971).

MORICK, H. (ed.), *Challenges to Empiricism* (Belmont, Calif.: Wadsworth, 1972).

PUTNAM, H., 'The Analytic and the Synthetic', in H. Feigl and G. Maxwell (eds.), *Minnesota Studies in the Philosophy of Science*, iii (Minneapolis: University of Minnesota Press, 1962), 358–97.

—— 'What Theories Are Not', in E. Nagel, P. Suppes, and A. Tarski (eds.), *Logic, Methodology and Philosophy of Science* (Stanford, Calif.: Stanford University Press, 1962), 240–51; repr. in Putnam, *Mathematics, Matter and Method, Philosophical Papers*, i (Cambridge: Cambridge University Press, 1975), 215–27.

QUINE, W., 'Two Dogmas of Empiricism', in *From a Logical Point of View* (Cambridge, Mass: Harvard University Press, 1953).

SCHAFFNER, K. F., 'Correspondence Rules', *Phil. Sci.* 36 (1969), 280–90.

SCRIVEN, M., 'Explanation and Prediction in Evolutionary Theory', *Science*, 130 (28 Aug. 1959), 477–82.

—— 'Explanations, Predictions, and Laws', in H. Feigl and G. Maxwell (eds.), *Minnesota Studies in the Philosophy of Science*, iii (Minneapolis: University of Minnesota Press, 1962), 170–230.

SELLARS, W., 'The Language of Theories', in B. Brody (ed.), *Readings in the Philosophy of Science*, 343–53.

SPECTOR, M., 'Models and Theories', *Br. J. Phil. Sci.* 16 (1965–6), 121–42.

TOULMIN, S., *Foresight and Understanding* (New York: Harper Torchbooks, 1961).

14. Theories of Scientific Progress

DILWORTH, C., *Scientific Progress* (Dordrecht: Reidel, 1981).

KORDIG, C., *The Justification of Scientific Change* (Dordrecht: Reidel, 1971).

KUHN, T. S., *The Essential Tension* (Chicago: University of Chicago Press, 1977).

—— *The Structure of Scientific Revolutions*, 2nd edn. (Chicago: University of Chicago Press, 1970).

LAKATOS, I., 'Falsification and the Methodology of Scientific Research Programmes', in I. Lakatos and A. Musgrave (eds.), *Criticism and the Growth of Knowledge*, (Cambridge: Cambridge University Press, 1970).

—— 'History of Science and Its Rational Reconstructions', in *Boston Studies in the Philosophy of Science*, viii (Dordrecht: Reidel, 1971), 91–136; this volume contains criticism of Lakatos's position by T. S. Kuhn, H. Feigl, R. J. Hall, and N. Koertge, and a reply by Lakatos.

—— and MUSGRAVE, A. (eds.), *Criticism and the Growth of Knowledge* (Cambridge: Cambridge University Press, 1970); includes essays critical of Kuhn's position by J. Watkins, S. Toulmin, L. P. Williams, K. Popper, M. Masterman, and P. Feyerabend, and a reply by Kuhn.

LAUDAN, L., *Progress and Its Problems* (Berkeley, Calif.: University of California Press, 1977).

MCMULLIN, E., 'The History and Philosophy of Science: A Taxonomy', in R. Stuewer (ed.), *Historical and Philosophical Perspectives of Science* (Minneapolis: University of Minnesota Press, 1970), 12–67.

MUSGRAVE, A., 'Kuhn's Second Thoughts', *Brit. J. Phil. Sci.* 22 (1971), 287–97.

SCHEFFLER, I., *Science and Subjectivity* (Indianapolis: Bobbs-Merrill, 1967); an attack on "subjective" alternatives to orthodoxy.

SHAPERE, D., 'The Structure of Scientific Revolutions', *Phil. Rev.* 73 (1964), 383–94.

15. Explanation, Causation, and Unification

DOWE, P., 'Wesley Salmon's Theory and the Conserved Quantity Theory', *Phil. Sci.* 59 (1992), 195–216.

GLYMOUR, G., 'Causal Inference and Causal Explanation', in R. McLaughlin (ed.), *What? Where? When? Why?* (Dordrecht: Reidel, 1982), 179–91.

HUMPHREYS, P., 'Scientific Explanation: The Causes, Some of the Causes, and Nothing But the Causes', in Kitcher and Salmon (eds.), *Scientific Explanation*, xiii. 283–306.

KITCHER, P., 'Explanatory Unification and the Causal Structure of the World', in Kitcher and Salmon (eds.), *Scientific Explanation*, xiii. 410–55.

SALMON, W., 'Causality: Production and Propagation', in P. D. Asquith and R. W. Giere (eds.), *PSA 1980*, ii (East Lansing Mich.: Philosophy of Science Association, 1981), 49–69.

—— *Scientific Explanation and the Causal Structure of the World* (Princeton, NJ: Princeton University Press, 1984).

—— 'Why Ask "Why"? An Inquiry Concerning Scientific Explanation', *Proc. Am. Phil. Soc.* 6 (1978), 685–701; repr. in Kourany (ed.), *Scientific Knowledge*, (Belmont, Calif.: Wadsworth, 1987), 51–64.

—— 'Four Decades of Scientific Explanation', in P. Kitcher and W. Salmon (eds.), *Scientific Explanation, Minnesota Studies in the Philosophy of Science*, xiii (Minneapolis: University of Minnesota Press, 1989); contains an extensive bibliography.

WOODWARD, J., 'The Causal Mechanical Model of Explanation', in Kitcher and W. Salmon (eds.), *Scientific Explanation*, xiii. 357–83.

16. Confirmation and Evidential Support

CAMPBELL, R., and VINCI, T., 'Novel Confirmation', *Brit. J. Phil. Sci.* 34 (1983), 315–41.

CHIHARA, C., 'Some Problems for Bayesian Confirmation Theory', *Brit. J. Phil. Sci.* 38 (1987), 551–60.

EARMAN, J. (ed.), *Testing Scientific Theories, Minnesota Studies in the Philosophy of Science*, x (Minneapolis: University of Minnesota Press, 1983); essays by P. Horwich, A. Edidin, R. Laymon, D. Garber, *et al.*

GARBER, D., 'Old Evidence and Logical Omniscience in Bayesian Confirmation Theory', in J. Earman (ed.), *Testing Scientific Theories*, (Minneapolis: University of Minnesota Press, 1983), 99–131.

GLYMOUR, C., *Theory and Evidence* (Princeton, NJ: Princeton University Press, 1980).

HOWSON, C., and URBACH, P., *Scientific Reasoning, The Bayesian Approach* (La Salle, Ill.: Open Court, 1989).

LAKATOS, I., 'Changes in the Problem of Inductive Logic', in Lakatos (ed.), *Inductive Logic* (Amsterdam: North-Holland, 1968), 375–90.

MILLER, R. W., *Fact and Method* (Princeton, NJ: Princeton University Press, 1987).

VAN FRAASSEN, B., 'Theory Comparison and Relevant Evidence', in Earman (ed.), *Testing Scientific Theories* (Minneapolis: University of Minnesota Press, 1983).

17. The Justification of Evaluative Standards

DOPPELT, G., 'Relativism and the Reticulational Model of Scientific Rationality', *Synthèse*, 69 (1986), 225–52.
—— 'The Naturalist Conception of Methodological Standards in Science: A Critique', *Phil. Sci.* 57 (1990), 1–19.
KUHN, T. S., 'Notes on Lakatos', in R. C. Buck and R. S. Cohen (eds.), *Boston Studies in the Philosophy of Science*, viii (Dordrecht: Reidel, 1971). 137–46.
LAUDAN, L., *Science and Values* (Berkeley: University of California Press, 1984).
—— 'Some Problems Facing Intuitionist Meta-Methodologies', *Synthèse*, 67 (1986), 115–29.
—— 'Progress or Rationality? The Prospects for a Normative Naturalism', *Amer. Phil. Quart.* 24 (1987), 19–31.
—— 'If It Ain't Broke, Don't Fix It', *Brit. J. Phil. Sci.* 40 (1989), 369–75; a reply to J. Worrall's 'The Value of a Fixed Methodology'.
—— 'Normative Naturalism', *Phil. Sci.* 57 (1990), 44–59.
LEPLIN, J., 'Renormalizing Epistemology', *Phil. Sci.* 57 (1990), 20–33.
LOSEE, J., *Philosophy of Science and Historical Enquiry* (Oxford: Clarendon Press, 1987).
ROSENBERG, A., 'Normative Naturalism and the Role of Philosophy, *Phil. Sci.* 57 (1990), 34–43.
SHAPERE, D., 'The Character of Scientific Change' in T. Nickles (ed.), *Scientific Discovery, Logic and Rationality* (Dordrecht: Reidel, 1980).
—— *Reason and the Search for Knowledge* (Dordrecht: Reidel, 1983).
WORRALL, J., 'The Value of a Fixed Methodology', *Brit. J. Phil. Sci.* 39 (1988), 263–75.
—— 'Fix It and Be Damned: A Reply to Laudan', *Brit. J. Phil. Sci.* 40 (1989), 376–88.

18. The Debate over Scientific Realism

BOYD, R., 'Scientific Realism and Naturalistic Epistemology', in P. D. Asquith and R. N. Giere (eds.), *PSA 1980*, ii (East Lansing, Mich.: Philosophy of Science Association, 1981), 613–39.
CARRIER, M., 'What Is Wrong with the Miracle Argument?' *Stud. Hist. Phil. Sci.* 22 (1991), 23–36.
CHURCHLAND, P. M., and HOOKER, C. A. (eds.), *Images of Science* (Chicago: University of Chicago Press, 1985); essays on van Fraassen's "Constructive Empiricism", by C. Glymour, I. Hacking, A. Musgrave, *et al.*
CLENDINNEN, C. J., 'Realism and the Underdetermination of Theory', *Synthèse*, 81 (1989), 63–90.

FINE, A., *The Shaky Game* (Chicago: University of Chicago Press, 1986), chs. 7–9.

HACKING, I., *Representing and Intervening* (Cambridge: Cambridge University Press, 1983).

HARRÉ, R., *Varieties of Realism* (Oxford: Blackwell, 1986).

LAUDAN, L., 'A Confutation of Convergent Realism', *Phil. Sci.* 48 (1981), repr. in J. Leplin (ed.), *Scientific Realism*, 218–49.

LEPLIN, J. (ed.), *Scientific Realism* (Berkeley, Calif.: University of California Press, 1984); essays by R. Boyd, A. Fine, C. Glymour, L. Laudan, B. van Fraassen, *et al.*

MUSGRAVE, A., 'The Ultimate Argument for Scientific Realism', in R. Nola (ed.), *Relativism and Realism in Science* (Dordrecht: Kluwer, 1988), 229–52.

ROUSE, J., 'Arguing for the Natural Ontological Attitude' (East Lansing, Mich.: Philosophy of Science Association, 1988), 294–301.

SMITH, P., *Realism and the Progress of Science* (Cambridge: Cambridge University Press, 1981).

SOBER, E., 'Constructive Empiricism and the Problem of Aboutness', *Brit. J. Phil. Sci.* 36 (1985), 11–18.

—— 'Contrastive Empiricism' in C. W. Savage (ed.), *Scientific Theories, Minnesota Studies in the Philosophy of Science*, xiv (Minneapolis: University of Minnesota Press, 1990), 392–409.

VAN FRAASSEN, B., *The Scientific Image* (Oxford: Clarendon Press, 1980).

WYLIE, A., 'Arguments for Scientific Realism: The Ascending Spiral', *Am. Phil. Quart.* 23 (1986), 287–97.

19. Descriptive Philosophies of Science

COHEN, L. J., 'Is the Progress of Science Evolutionary?' *Brit. J. Phil Sci.* 24 (1973), 41–61.

HOLTON, G., *The Scientific Imagination* (Cambridge: Cambridge University Press, 1978).

—— 'Do Scientists Need a Philosophy?' *Times Literary Supplement*, 2 Nov. 1984, 1232–3.

—— *Thematic Origins of Scientific Thought*, rev. edn. (Cambridge, Mass.: Harvard University Press, 1988).

HULL, D. L., *The Metaphysics of Evolution* (Albany, NY: SUNY Press, 1989).

—— *Science as a Process* (Chicago: University of Chicago Press, 1988).

TOULMIN, S., *Human Understanding*, i (Oxford: Clarendon Press, 1972).

—— 'Rationality and Scientific Discovery', in R. S. Cohen and M. Wartofsky (eds.), *Boston Studies in the Philosophy of Science*, xx (Dordrecht: Reidel, 1974), 387–406.

Index of Proper Names

Entries in lists of notes are indexed in italics

Index of Subjects

OXFORD

MORE OXFORD PAPERBACKS

This book is just one of nearly 1000 Oxford Paperbacks currently in print. If you would like details of other Oxford Paperbacks, including titles in the World's Classics, Oxford Reference, Oxford Books, OPUS, Past Masters, Oxford Authors, and Oxford Shakespeare series, please write to:

UK and Europe: Oxford Paperbacks Publicity Manager, Arts and Reference Publicity Department, Oxford University Press, Walton Street, Oxford OX2 6DP.

Customers in UK and Europe will find Oxford Paperbacks available in all good bookshops. But in case of difficulty please send orders to the Cash-with-Order Department, Oxford University Press Distribution Services, Saxon Way West, Corby, Northants NN18 9ES. Tel: 0536 741519; Fax: 0536 746337. Please send a cheque for the total cost of the books, plus £1.75 postage and packing for orders under £20; £2.75 for orders over £20. Customers outside the UK should add 10% of the cost of the books for postage and packing.

USA: Oxford Paperbacks Marketing Manager, Oxford University Press, Inc., 200 Madison Avenue, New York, N.Y. 10016.

Canada: Trade Department, Oxford University Press, 70 Wynford Drive, Don Mills, Ontario M3C 1J9.

Australia: Trade Marketing Manager, Oxford University Press, G.P.O. Box 2784Y, Melbourne 3001, Victoria.

South Africa: Oxford University Press, P.O. Box 1141, Cape Town 8000.

PHILOSOPHY IN OXFORD PAPERBACKS
THE GREAT PHILOSOPHERS
Bryan Magee

Beginning with the death of Socrates in 399, and following the story through the centuries to recent figures such as Bertrand Russell and Wittgenstein, Bryan Magee and fifteen contemporary writers and philosophers provide an accessible and exciting introduction to Western philosophy and its greatest thinkers.

Bryan Magee in conversation with:

A. J. Ayer	John Passmore
Michael Ayers	Anthony Quinton
Miles Burnyeat	John Searle
Frederick Copleston	Peter Singer
Hubert Dreyfus	J. P. Stern
Anthony Kenny	Geoffrey Warnock
Sidney Morgenbesser	Bernard Williams
Martha Nussbaum	

'Magee is to be congratulated . . . anyone who sees the programmes or reads the book will be left in no danger of believing philosophical thinking is unpractical and uninteresting.' Ronald Hayman, *Times Educational Supplement*

'one of the liveliest, fast-paced introductions to philosophy, ancient and modern that one could wish for' *Universe*

Oxford Reference

The Oxford Reference series offers authoritative and up-to-date reference books in paperback across a wide range of topics.

ILLUSTRATED HISTORIES IN
OXFORD PAPERBACKS

THE OXFORD ILLUSTRATED HISTORY
OF ENGLISH LITERATURE

Edited by Pat Rogers

Britain possesses a literary heritage which is almost
unrivalled in the Western world. In this volume, the
richness, diversity, and continuity of that tradition
are explored by a group of Britain's foremost literary scholars.

Chapter by chapter the authors trace the history
of English literature, from its first stirrings in Anglo-
Saxon poetry to the present day. At its heart towers
the figure of Shakespeare, who is accorded a special
chapter to himself. Other major figures such as
Chaucer, Milton, Donne, Wordsworth, Dickens,
Eliot, and Auden are treated in depth, and the story
is brought up to date with discussion of living
authors such as Seamus Heaney and Edward Bond.

'[a] lovely volume . . . put in your thumb and pull
out plums' Michael Foot

'scholarly and enthusiastic people have written in-
spiring essays that induce an eagerness in their read-
ers to return to the writers they admire' *Economist*